TERRORISM, HOT SPOTS AND CONFLICT-RELATED ISSUES

CHEMICAL FACILITY ANTI-TERRORISM STANDARDS

OVERVIEW AND IMPLEMENTATION ISSUES

TERRORISM, HOT SPOTS AND CONFLICT-RELATED ISSUES

Additional books in this series can be found on Nova's website under the Series tab.

Additional e-books in this series can be found on Nova's website under the e-book tab.

TERRORISM, HOT SPOTS AND CONFLICT-RELATED ISSUES

CHEMICAL FACILITY ANTI-TERRORISM STANDARDS

OVERVIEW AND IMPLEMENTATION ISSUES

JOANNE NELSON
EDITOR

New York

NOTICE TO THE READER

Library of Congress Cataloging-in-Publication Data

ISBN: 978-1-63484-465-9

Published by Nova Science Publishers, Inc. † New York

CONTENTS

Preface vii

Chapter 1 Implementation of Chemical Facility Anti-Terrorism Standards (CFATS): Issues for Congress 1
Dana A. Shea

Chapter 2 Critical Infrastructure Protection: DHS Action Needed to Verify Some Chemical Facility Information and Manage Compliance Process 25
United States Government Accountability Office

Index 71

PREFACE

The potential harm to public health and the environment from a large release of hazardous chemicals has long concerned congressional policymakers. Since 2006, Congress has authorized the Department of Homeland Security (DHS) to regulate chemical facilities for security purposes. To this end, the Chemical Facility Anti-Terrorism Standards (CFATS) regulates security at high-risk facilities possessing more than certain amounts of one or more chemicals of interest. Facilities possessing more than the specified amount must register with DHS through this program and perform security-related activities. This book provides information regarding the pace of DHS CFATS implementation and attempts to identify potential sources of challenges for DHS. Moreover, this book addresses the extent to which DHS has categorized facilities as subject to the CFATS regulation, and approved site security plans and conducted compliance inspections.

In: Chemical Facility Anti-Terrorism Standards ISBN: 978-1-63484-465-9
Editor: Joanne Nelson

Chapter 1

IMPLEMENTATION OF CHEMICAL FACILITY ANTI-TERRORISM STANDARDS (CFATS): ISSUES FOR CONGRESS*

Dana A. Shea

SUMMARY

The Department of Homeland Security (DHS) implements the Chemical Facility Anti-Terrorism Standards (CFATS) regulations, which regulate security at high-risk facilities possessing more than certain amounts of one or more chemicals of interest. Facilities possessing more than the specified amount must register with DHS through this program (a process known as the Top-Screen) and perform security-related activities. The DHS identifies a subset of high-risk chemical facilities from among those that register. These high-risk chemical facilities must submit a security vulnerability assessment, which DHS uses to confirm their high-risk designation, and a site security plan, which DHS then reviews and authorizes. The DHS also inspects and approves high-risk chemical facilities for adherence to their submitted site security plans. It also later inspects for compliance with these plans following DHS approval. The DHS regulates approximately 3,600 facilities under this program and is in the process of implementing the regulatory

* This is an edited, reformatted and augmented version of a Congressional Research Service publication R43346, prepared for Members and Committees of Congress, dated February 25, 2015.

requirements for security vulnerability assessment, site security planning, and inspection.

The DHS has had challenges meeting its own projections and congressional expectations regarding program performance, raising questions about its ability to achieve steady-state regulatory implementation. As of April 2014, DHS had made final risk assignments to 3,316 of the approximately 4,200 facilities then regulated. The DHS reported in December 2014 that it now regulates approximately 3,600 facilities, approximately 2,700 of which have final risk assignments. The DHS *authorizes* a site security plan when the submitted plan satisfies CFATS requirements. Following a successful authorization inspection, DHS *approves* the site security plan. As of February 2014, DHS had authorized 2,646 site security plans; conducted 1,969 authorization inspections; and approved 1,526 site security plans. Between February 2014 and February 2015, DHS authorized an average of 132 and approved an average of 85 site security plans per month. The DHS has generally increased its average authorization and approval rate over time; between August 2014 and February 2015, DHS authorized an average of 135 and approved an average of 93 site security plans per month. Between January 2015 and February 2015, DHS authorized 120 and approved 80 site security plans.

This report analyzes data from a variety of DHS presentations, testimony, and other sources to present a historical overview of program performance to date. It identifies an ongoing gap between the number of facilities that have received final risk tier assignments and the total number of regulated facilities. This makes it appear likely that DHS will not have inspected or approved site security plans for some portion of the regulated facility universe for at least several years. In addition, the current rates of authorization of site security plans, authorization inspection, and approval of site security plans make it appear likely that DHS will not have completed implementation for the initial facilities before the date when it will potentially need to begin reinspecting already approved facilities. With the onset of compliance inspections, congressional policymakers may have further questions about the ability of the CFATS program to meet congressional expectations regarding timeliness.

INTRODUCTION

The chemical industry is one of the United States' largest manufacturing industries, directly employing more than 790,000 workers. The chemical industry serves both a domestic and a global market, and other industrial sectors rely on chemical products as essential to their business. The potential harm to public health and the environment from a large release of hazardous chemicals has long concerned congressional policymakers. Accidental chemical releases spurred environmental and other legislative proposals to reduce the risk of chemical accidents in the United States. While congressional interest in the security of chemical facilities predates the September 11, 2001, attacks, they prompted renewed congressional attention to the potential security risks posed by hazardous chemicals.

In 2006, Congress authorized the Department of Homeland Security (DHS) to regulate chemical facilities for security purposes.[1] In 2014, Congress reaffirmed this authority with some modifications.[2] In 2007, DHS issued interim final regulations establishing risk-based performance standards known as the Chemical Facility Anti-Terrorism Standards (CFATS). These regulations provided a process whereby facilities would submit information and security plans to DHS, DHS would review and approve these plans, facilities would implement them, and DHS would inspect their implementation. These regulations continue in force so long as they do not conflict with the new statutory authority enacted in 2014.

The DHS has historically experienced challenges in implementing the CFATS regulations. It has not met its own expected timelines or milestones to begin[3] or complete facility inspections.[4] As of February 2015, DHS had inspected and approved 1,526 site security plans, covering approximately 43% of the regulated facilities.

This report provides information regarding the pace of DHS CFATS implementation and attempts to identify potential sources of challenges for DHS. It does not address broader issues related to chemical facility security; for such information and analysis, see CRS Report R42918, *Chemical Facility Security: Issues and Options for the 113th Congress*, by Dana A. Shea.

Statute and Regulation

In 2006, Congress authorized the Department of Homeland Security (DHS) to issue interim final regulations establishing risk-based performance standards for chemical facility security and requiring facilities to develop vulnerability assessments and to develop and implement site security plans.[5] Among other provisions, the statute directs DHS to allow regulated entities to employ combinations of security measures to meet risk-based performance standards,[6] and it specifics that these regulations are to apply only to those chemical facilities that the Secretary determines present high levels of security risk.

The law contains a high-level framework for its implementation. The Secretary must review and approve the required vulnerability assessment, site security plan, and its implementation at each facility. The Secretary must audit and inspect chemical facilities and determine regulatory compliance. If the Secretary finds that a facility is not in compliance, the Secretary must write to the facility explaining the deficiencies found, provide an opportunity for the facility to consult with the Secretary, and issue an order to the facility to comply by a specified date. If the facility continues to be out of compliance, the Secretary may impose a fine and, eventually, order the facility to cease operation.

In 2007, the DHS issued an interim final rule and created the Chemical Facility Anti-Terrorism Standards (CFATS) program.[7] Much of the rule arises from the Secretary's discretion and DHS interpretation of legislative intent and was not explicitly detailed by the law.

Under the interim final rule, the Secretary of Homeland Security determined which chemical facilities must meet regulatory security requirements. Facilities with greater than specified quantities of any of 322 chemicals of interest must submit information to DHS, via a process known as the "Top-Screen," so that DHS can preliminarily determine each facility's risk status. Only those DHS deems high-risk are subject to the CFATS program and regulation.

The DHS assigns all high-risk facilities one of four risk-based tiers. The DHS established different performance-based requirements for facilities assigned to each risk-based tier. Facilities in higher risk tiers must meet more stringent performance-based requirements (Tier 1 being the most stringent and Tier 4 being the least stringent).

All high-risk facilities must develop a vulnerability assessment and an effective security plan, submit these documents to DHS, and implement their security plan. The vulnerability assessment serves two purposes under the interim final rule. One is to determine or confirm the placement of the facility in a risk-based tier. The other is to provide a baseline against which to assess the effectiveness of actions identified in the site security plan.

Each site security plan must describe how activities in the plan address issues identified in the facility vulnerability assessment. Additionally, the site security plan must address preparations for and deterrents against specific modes of potential terrorist attack, as applicable and identified by DHS. The site security plan must also describe how the activities taken by the facility meet the risk-based performance standards provided by DHS.

High-risk facilities may develop vulnerability assessments and site security plans using alternative security programs established by a third party, such as an industry organization program,[8] so long as they meet the tiered, performance-based requirements of the interim final rule. The Secretary may disapprove submitted vulnerability assessments or site security plans that fail to meet DHS standards but not on the basis of the presence or absence of a specific measure. In the case of disapproval, DHS must identify in writing those areas of the assessment and plan that need improvement. Chemical facilities may request further adjudication by DHS of site security plan disapprovals.

Once facilities have complied with the regulation by submitting security documentation and receiving approval of their plans, they are to implement their site security plans. The DHS then monitors facility compliance with these site security plans. The DHS indicated in 2007 that it would generally require Tier 1 and Tier 2 facilities to update their Top-Screen, security vulnerability assessment, and site security plan every two years and Tier 3 and Tier 4 facilities to do so every three years.[9]

In 2014, Congress provided new authority to DHS to regulate security at chemical facilities.[10] The reauthorization generally supports the existing regulatory structure and expressly allows DHS to continue using the existing regulations unless they conflict with the new authority. The new statute also contains provisions, such as an alternative self-certification program, that likely require action on DHS to implement. The DHS has not yet issued new regulations to implement these provisions.

IMPLEMENTATION ANALYSIS

In effect, the CFATS regulation and its underlying statute create a multistep, iterative process of facility information submission followed by DHS approval. Each step is iterative since both the facility, which may revise its submissions, and DHS, which may disapprove the submitted documents, may cause the other participant to repeat the information exchange. See *Figure 1*.

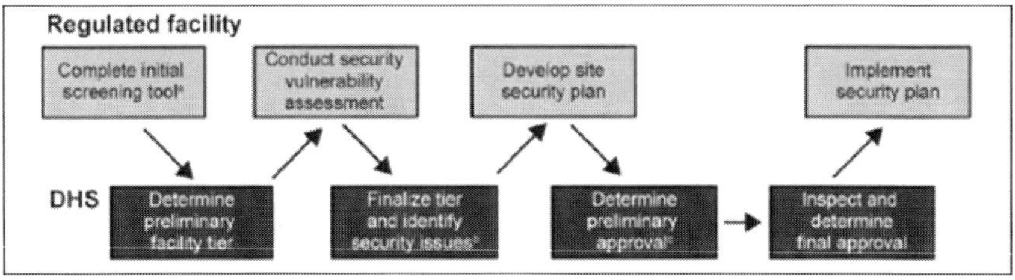

Source: Government Accountability Office, Critical Infrastructure Protection: DHS Needs to Improve Its Risk Assessments and Outreach for Chemical Facilities, GAO-13-801T, August 1, 2013.

Notes: The DHS also performs compliance inspections after a facility implements its site security plan. During each of the above steps, the facility, DHS, or both, may request additional information or propose to change a determination.

[a] Facilities are to submit an initial screening tool (the Top-Screen) that provides basic information about the facilities and the chemicals they possess.

[b] This step includes determining if a facility has high risk, and if so, DHS assigns a tier and identifies security issues.

[c] At this stage, if requirements are satisfied, DHS issues a letter of authorization for the facility's plan.

Figure 1. Implementation Steps in the CFATS Process.

Facilities possessing a chemical of interest in quantities above the screening threshold quantity submit a Top Screen. Top Screen results assist the CFATS program in determining whether a facility presents a high-level security risk. After processing a Top Screen, the CFATS Program assigns the facility a preliminary tier or determines that the facility does not meet the criteria for CFATS regulation. When a facility receives a preliminary tier assignment notification, it must prepare and submit a Security Vulnerability Assessment (SVA) to DHS. After reviewing the SVA, the CFATS Program determines a facility's final tier assignment or that the facility is not high-risk. When the facility receives a final tier assignment, it must develop and submit a Site Security Plan (SSP) to DHS.

The Infrastructure Security Compliance Division (ISCD) in the DHS National Protection and Programs Directorate (NPPD) reviews the SSP to preliminarily determine if it satisfies the applicable risk-based performance standards. This process typically involves discussions between ISCD staff and the facility. It often requires the facility to submit additional information to ISCD and revise the SSP before ISCD can complete its initial review and issue the facility a letter of authorization for the SSP.

After issuing a letter of authorization, ISCD conducts a comprehensive and detailed authorization inspection. The ISCD reviews the inspection results, as well as any further revisions that the facility may make to the SSP. It then makes a final determination as to whether the facility's SSP satisfies the applicable risk-based performance standards. If so, ISCD issues a letter of approval, and the facility must implement the applicable provisions of the SSP. If ISCD determines that the SSP does not satisfy the applicable risk-based performance standards, ISCD may issue a notice of deficiency and require the facility to resubmit a sufficient SSP. If the facility fails to do so, ISCD may disapprove the SSP. Following the authorization inspection, facilities are generally granted an additional 45 days to make any necessary modifications to their SSP. ISCD will then review the resubmitted SSP and make a final determination as to whether the SSP warrants the issuance of a letter of approval. In some cases, the facility may require a technical consultation and another opportunity to revise its SSP.

Over time, the DHS has attempted to develop a consistent nomenclature for its review and inspection process.[11] The DHS *authorizes* an SSP (issuing the facility a *letter of authorization*) when the submitted SSP is satisfactory under CFATS.[12] The DHS conducts an *authorization inspection* of a facility with an authorized SSP to compare the authorized SSP to the conditions of the facility. Following a successful authorization inspection, the DHS *approves* the SSP (issuing the facility a *letter of approval*). At a later date, expected by DHS to be one year after approval of the SSP, the DHS will conduct a *compliance inspection* of a facility to determine whether the facility has fully implemented its approved SSP.[13] Compliance inspections then occur on a periodic basis depending on the risk tier to which the facility is assigned.

The following sections analyze DHS data to identify underlying challenges for the following steps: the process of DHS assigning a facility to a risk tier; the rate at which DHS authorizes site security plans; the rate at which DHS inspects and approves authorized site security plans; and the occurrence of compliance inspections.

Chemical Facility Tier Assignment

In the CFATS process, all facilities possessing chemicals of interest above a certain threshold amount must report to DHS through the Top-Screen process. At the inception of the program, DHS received approximately 29,000 Top-Screen submissions and preliminarily identified approximately 7,000 facilities as high-risk and therefore subject to CFATS regulation. Since that time, DHS has received a cumulative total of more than 48,000 Top-Screen submissions from over 36,000 facilities.[14] Some of this increase is attributable to Top-Screen submissions from facilities that had never previously submitted one. Such facilities include those that have increased their holdings of chemicals of interest above the specified threshold and facilities that had not previously complied with the regulation. Some of these new submissions will be from facilities that are not high-risk. The remaining additional Top-Screens represent resubmissions from facilities with new facility information, modified activities, or changed chemical holdings. Based on this information, DHS may reassess whether facilities should be assigned to a different high-risk tier or removed from the program altogether.[15]

Despite the increase in cumulative Top-Screen submissions, the number of facilities meeting the DHS criteria for high risk has fallen over time to approximately 3,600. See *Figure 2*. This decrease represents facilities that have reduced their holdings of chemicals of interest below threshold levels, which could be caused by voluntary reductions or facility closures. The DHS has stated that it no longer considers as having high risk more than 3,000 facilities that voluntarily removed, reduced, or modified their holdings of chemicals of interest.[16]

An appreciable delay may be occurring between the submission of Top-Screen information and DHS identifying a facility as preliminarily high-risk. In 2013, the DHS Office of Inspector General (OIG) analyzed DHS responses to Tier 1 and Tier 2 Top-Screen submissions. It found that the average time to receive a tier assignment was 4.8 months, with the longest observed time being 12 months.[17] This delay might cause the number of Top-Screen submissions to rise without a commensurate increase in the number of regulated facilities until such time as DHS responds to the new submissions.

The decrease in the number of regulated facilities results from more facilities leaving the program than entering the program. To the extent that the additional Top-Screen submissions reflect information from new facilities, then a commensurately greater number of facilities must be leaving the CFATS program.

The increasing number of Top-Screen submissions combined with the decreasing number of regulated facilities suggests that some fraction of the existing facility universe is submitting multiple Top-Screens. The filing of multiple Top-Screen submissions may pose a challenge to DHS, since it may trigger a review of the facility's risk tier, related documentation, and required risk-based performance standards.

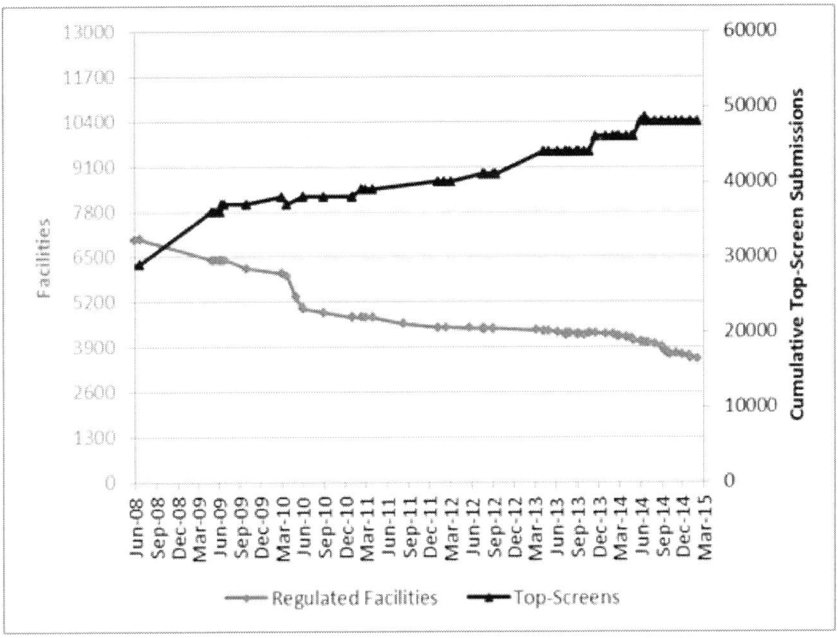

Source: CRS analysis of various DHS presentations, testimony, and other sources.

Notes: Total regulated facilities plotted against the left axis; Cumulative Top-Screen submissions plotted against the right axis. The DHS generally expresses the number of Top-Screen submissions as "more than" a certain amount and reports in thousands. Thus, the number of Top-Screen submissions should be taken as a lower bound.

Figure 2. Regulated Facilities and Top-Screen Submissions. (June 2008 through February 2015).

Once a facility submits Top-Screen data, DHS assigns a preliminary risk tier to the facility. The DHS assigns a final risk tier after reviewing the facility's security vulnerability assessment. As shown in *Figure 3*, DHS has provided final tier assignments to most facilities. Since mid-2010, between 14% and 25% of high-risk facilities at any given time lack a final tier assignment.

Analysis of these data on a tier basis reveals that 63% of the facilities lacking a final tier assignment are in Tier 4.[18]

The reasons why facilities lack a final tier assignment are likely to be a combination of several factors:

- Facilities newly joining the CFATS program are assigned a preliminary tier. Some of these new submissions may be from high-risk facilities.
- Facilities with incomplete or insufficient security vulnerability assessments do not receive final tier assignments until DHS and the facility resolved outstanding issues.
- Facilities changing the amount of chemicals of interest on site may request a new tiering determination. During this process, a facility may be identified as preliminarily tiered rather than having a final tier assignment.
- A facility contesting its determination as a high-risk facility or its assignment to a specific risk tier may not receive a final tier assignment.
- Some facilities are in categories for which DHS has indefinitely extended the timeline for compliance. For example, DHS indefinitely extended the timeline for compliance for facilities with aboveground gasoline storage tanks, including facilities (such as petroleum refineries) that may possess other chemicals that trigger the Top-Screen requirement. Approximately 405 such facilities are preliminarily high-risk.[19]

A key issue for oversight is how DHS enforces CFATS on facilities receiving a preliminary tier assignment. The DHS requires a facility to begin the site security planning process only after the final assignment to a risk tier. Consequently, DHS oversight of security at facilities awaiting final risk-tier assignment seems likely to be delayed relative to other facilities. In addition, the fraction of facilities receiving a final tier assignment has trended slightly downward over time. This may reflect the dynamic nature of the pool of CFATS-regulated facilities with facilities joining, leaving, and being retiered within the CFATS program, or it may suggest that DHS is not resolving the gap between total facilities and facilities with a final tier.

Site Security Plan Authorization

Regulated facilities must submit site security plans (SSPs) that meet the risk-based performance standards applicable to their final risk tier. The DHS reviews and issues a letter of authorization to facilities submitting a satisfactory SSP. The DHS has experienced challenges in developing an effective and efficient authorization process and has engaged in internal review to improve that process.

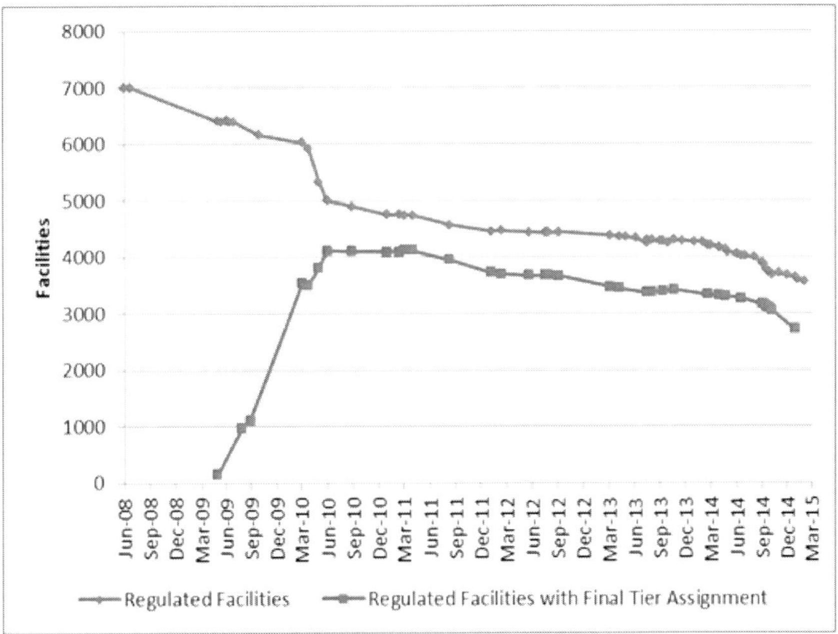

Source: CRS analysis of various DHS presentations, testimony, and other sources.

Notes: The gap between the number of facilities and the number with final tier assignments is facilities with preliminary tier assignments.

Figure 3. High-Risk Facilities Assigned Final Risk Tiers. (June 2008 through February 2015).

Figure 4 shows the number of authorized site security plans. The DHS authorized its first site security plan on May 28, 2010.[20] Between then and August 2012, when DHS implemented a new integrated SSP review process, DHS authorized approximately 60 site security plans. Since August 2012, the rate of site security plan authorization has increased. The DHS has authorized a total of 2,646 SSPs as of February 2015. Facilities in all risk tiers have received authorizations, though DHS has focused on Tier 1 and Tier 2 facilities.

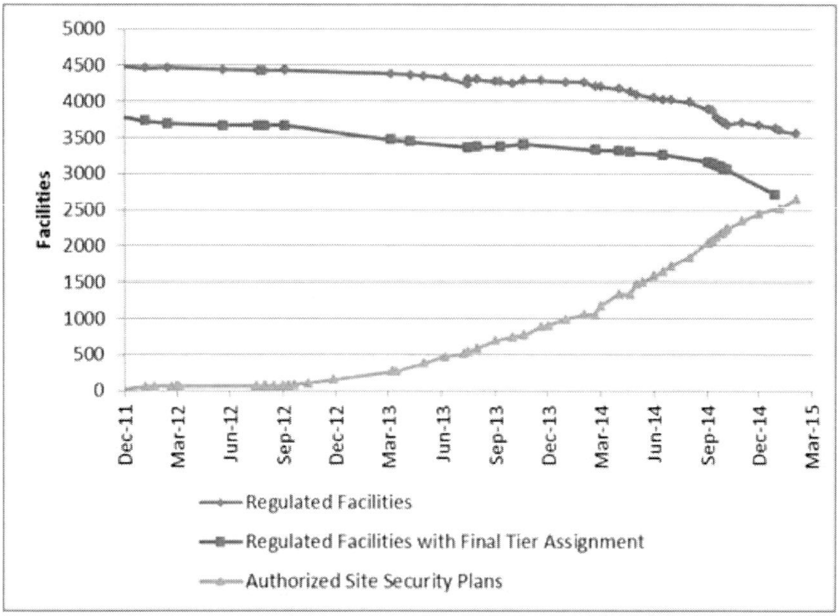

Source: CRS analysis of various DHS presentations, testimony, and other sources.

Notes: The decline in the number of authorized site security plans in February 2014 reflects removal by DHS of authorized site security plans for facilities no longer regulated under CFATS.

Figure 4. Authorized Site Security Plans. (January 2012 to February 2015).

The current authorization rate is significantly higher than it was previously, and 74% of all regulated facilities (97% of facilities having received a final tier assignment as of December 2014) have an authorized SSP as of February 2015. The DHS has authorized most facilities that have submitted an SSP.[21] Comparison of the current rate of authorization against that necessary to authorize all SSPs suggests that DHS may finish authorization of the existing SSPs in less than a year. See *Figure 5*.

Figure 5 shows trajectories required to complete SSP authorization within 1, 2, 5, or 10 years. Assuming that the total number of regulated facilities remains unchanged over time, DHS would need to authorize an average of 76 site security plans monthly to authorize a number of SSPs equal to the number of facilities within 1 year, an average of 38 monthly to authorize all within 2 years, an average of 15 monthly to authorize all within 5 years, and an average of 8 monthly to authorize all within 10 years. As a point of reference, DHS authorized an average of 132 SSPs per month from February 2014 to February

2015.[22] The DHS has generally increased its average authorization rate over time; between August 2014 and February 2015, DHS authorized an average of 135 SSPs per month.[23] Between January 2015 and February 2015, DHS authorized 120 SSPs.[24] See *Table 1*.

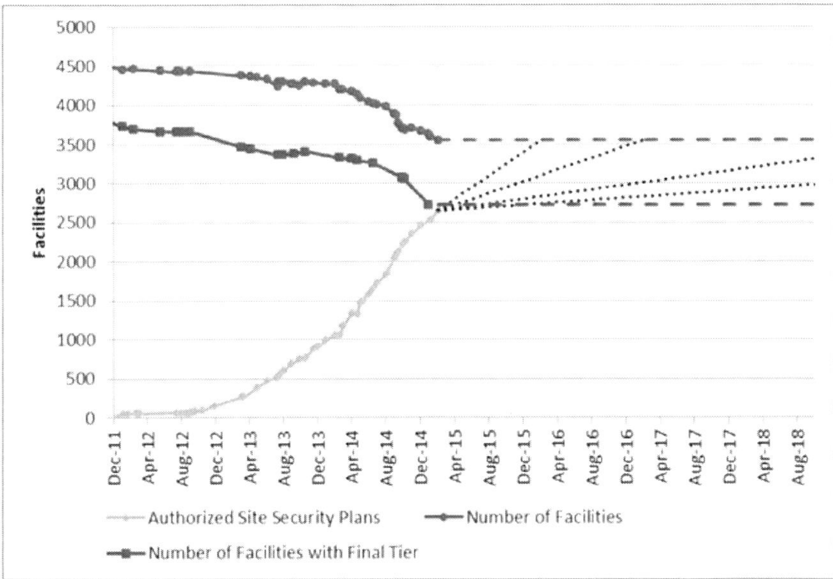

Source: CRS analysis of various DHS presentations, testimony, and other sources.

Notes: Dotted lines indicate the trajectories required to complete site security plan authorization within 1, 2, 5, or 10 years. Projections assume a constant rate of authorization, the authorization of SSPs for all facilities (including those currently without a final tier), no change in the total number of facilities, and that each site security plan corresponds to a single regulated facility.

Figure 5. Projection for Completion of SSP Authorization in 1, 2, 5, and 10 Years.

The past rate of SSP authorization may not reflect future performance. The DHS may be able to increase the rate of SSP authorization. The DHS has so far mostly been assessing SSPs for Tier 1 and Tier 2 facilities; SSPs for Tier 3 and Tier 4 facilities may be simpler. As DHS obtains greater experience in authorizing SSPs, the process may become more efficient. Conversely, DHS may have difficulty maintaining the current rate of SSP authorizations if future SSPs have lower quality or pose particular challenges or if other activities compete for ISCD resources. Note that the projections reported from the above model assume that no facilities join or leave the CFATS program and that each facility requires only a single authorized site security plan.

Table 1. Projected Required Site Security Plan Authorization Rate

(average monthly rate)

Timeframe	Monthly Number of Authorized Site Security Plans
Actual Recent Rate (over 12 months, see text)	132
Actual Recent Rate (over 6 months, see text)	135
Actual Recent Rate (over 1 month, see text)	120
Rate Required to Complete in 1 Year	76
Rate Required to Complete in 2 Years	38
Rate Required to Complete in 5 Years	15
Rate Required to Complete in 10 Years	8

Source: CRS analysis of various DHS presentations, testimony, and other sources.

Notes: Projections assume a constant rate of authorization, the authorization of SSPs for all facilities (including those currently without a final tier), no change in the total number of facilities, and that each site security plan corresponds to a single regulated facility.

A more pressing challenge may be that most facilities without an authorized SSP also lack a final tier assignment. As such, these facilities are not required to yet submit an SSP (see *Figure 1*). Consequently, resolving the security of these facilities may take additional time, possibly longer than expected from the current rate of authorization.

Authorization Inspections and SSP Approvals

Following authorization of a site security plan, DHS inspects the facility to determine its adherence to its submitted SSP. Once DHS has confirmed such compliance, DHS approves the SSP and provides the facility with a letter of approval. The DHS began performing authorization inspections in July 2010 but, following an internal review, suspended its inspection process pending revisions. The DHS resumed authorization inspections in July 2012 and had completed more than 100 authorization inspections by April 2013. The DHS did not approve site security plans at the same rate as it conducted inspections. See *Figure 6*.

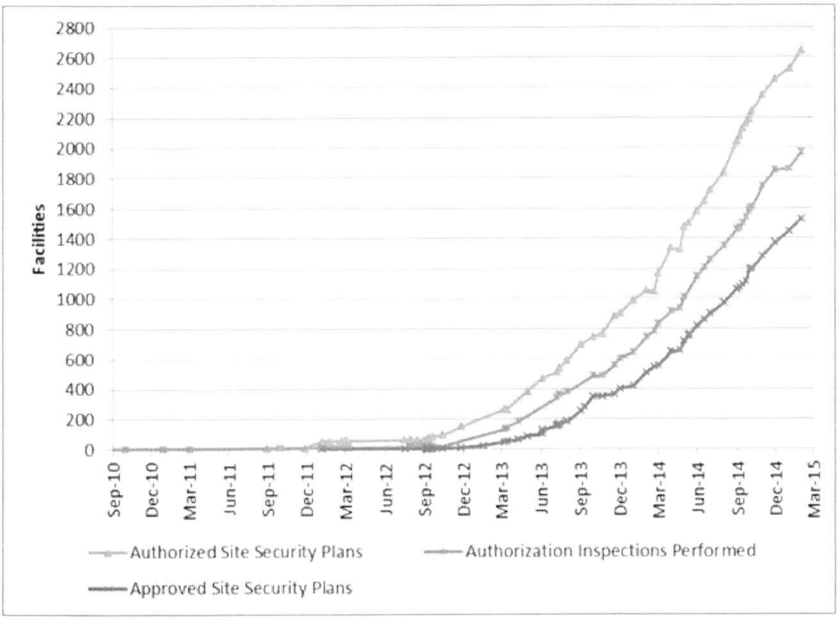

Source: CRS analysis of various DHS presentations, testimony, and other sources.

Note: The decline in the number of authorized site security plans in February 2014 reflects removal by DHS of authorized site security plans for facilities no longer regulated under CFATS.

Figure 6. Authorized SSPs, Authorization Inspections, and Approved SSPs. (October 2010 to February 2015).

The number of authorized SSPs and the number of authorization inspections appear to roughly parallel each other. This implies that the authorization inspection rate is roughly commensurate with the rate of SSP authorization. Assuming a single inspection at each facility, the data do not suggest that an additional backlog of authorized SSPs waiting for authorization inspections is likely to build up. If the SSP authorization rate were to increase relative to the rate of authorization inspections, an additional backlog of authorized SSPs could develop. That effect might be addressed through increased inspection capability.

Similarly, the rate of authorization inspections roughly parallels the rate of SSP approval. The number of authorization inspections is greater than that of approved SSPs. One possibility is that not all authorization inspections lead to an approval. Some facilities may be disapproved, while others may require multiple authorization inspections to resolve challenges identified in the inspection process. Another possibility is that the process of issuing an approval following an authorization inspection takes time; DHS has reported that a facility may be granted up to 45 days to amend a site security plan following an authorization inspection.[25]

As of February 2015, 57% of all regulated facilities (44% of facilities with a final tier assignment as of December 2014) lack an approved SSP.[26] Comparison of the current rate of approval against that necessary to approve all SSPs indicates that it will likely take more than a year for DHS to finish inspecting and approving the existing SSPs. See *Figure 7*.

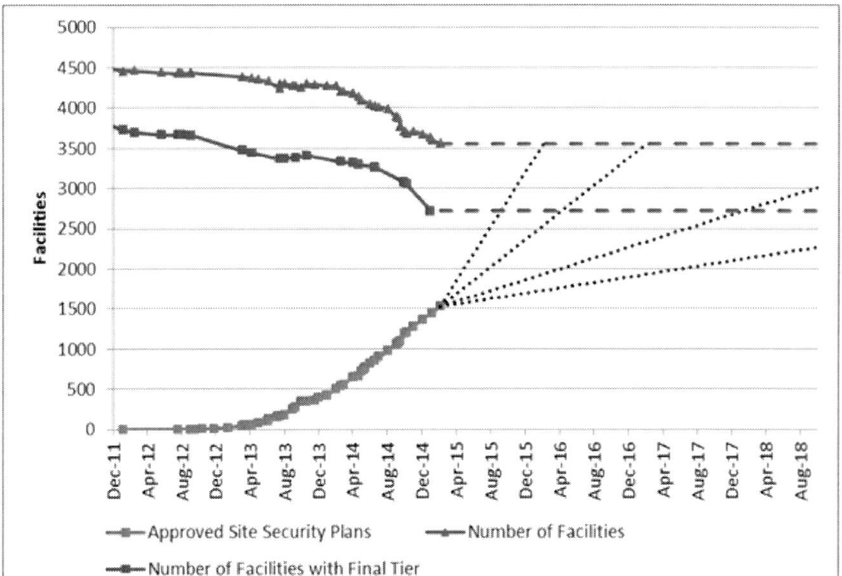

Source: CRS analysis of various DHS presentations, testimony, and other sources.

Notes: Dotted lines indicate the trajectories required to complete site security plan authorization within 1, 2, 5, or 10 years. Projections assume a constant rate of authorization, the authorization of SSPs for all facilities (including those currently without a final tier), no change in the total number of facilities, and that each site security plan corresponds to a single regulated facility.

Figure 7. Projection for Completion of SSP Approval in 1, 2, 5, and 10 Years.

If the total number of regulated facilities remained unchanged over time, DHS would need to approve an average of 169 site security plans monthly to approve a number of SSPs equal to the number of facilities within 1 year, an average of 84 monthly to approve all within 2 years, an average of 34 monthly to approve all within 5 years, and an average of 17 monthly to approve all within 10 years. As a point of reference, DHS approved an average of 85 site security plans monthly between February 2014 and February 2015.[27] The DHS has generally increased this rate over time; between August 2014 and February 2015, DHS approved an average of 93 SSPs per month.[28] Between January 2015 and February 2015, DHS approved 80 SSPs.[29] The Government Accountability Office estimated in April 2013 that it could take DHS from seven to nine years to complete reviews and approvals for SSPs then submitted to DHS from facilities with a final risk tier.[30] The rate of approvals has increased since that GAO analysis. See *Table 2*.

Table 2. Projected Required Site Security Plan Approval Rate

(average monthly rate)

Timeframe	Monthly Number of Approved Site Security Plans
Actual Recent Rate (over 12 months, see text)	85
Actual Recent Rate (over 6 months, see text)	93
Actual Recent Rate (over 1 month, see text)	80
Rate Required to Complete in 1 Year	169
Rate Required to Complete in 2 Years	84
Rate Required to Complete in 5 Years	34
Rate Required to Complete in 10 Years	17

Source: CRS analysis of various DHS presentations, testimony, and other sources.

Notes: Projections assume a constant rate of authorization, the authorization of SSPs for all facilities (including those currently without a final tier), no change in the total number of facilities, and that each site security plan corresponds to a single regulated facility.

The past rate of SSP approval may not reflect future performance. The DHS may be able to increase the rate of SSP approval. The DHS has so far mostly been inspecting Tier 1 and Tier 2 facilities; Tier 3 and Tier 4 facilities may be simpler. As DHS gets more experience in inspecting facilities and issuing SSP approvals, the process may become more efficient. Conversely, DHS may have difficulty maintaining the current rate of SSP approvals if future SSPs have lower quality or pose particular challenges. Note that the projections above assume that no facilities join or leave the CFATS program and that each facility requires only a single approved site security plan.

Compliance Inspections

Following approval of a facility site security plan, DHS inspects the facility for compliance. According to DHS, compliance inspections started in September 2013. As of May 2014, DHS had performed 31 compliance inspections[31] and 61 compliance inspections as of December 2014.[32] In April 2013, GAO estimated that full regulatory implementation, including compliance inspection, for existing facilities would require 8 to 10 years.[33] Insufficient data are available to assess whether the rate of compliance inspection has increased since the GAO analysis. The rate of compliance inspections does not meet that of all facility approvals from one year prior.[34]

An additional component of future compliance involves the resubmission of updated Top-Screen information, and if so requested, security vulnerability assessments and site security plans. According to DHS's explanation of the CFATS regulation, "the Department will require facilities in Tiers 1 and 2 to update their Top-Screen, SVA, and SSP every two years, and facilities in Tiers 3 and 4 to update their Top-Screen, SVA, and SSP every three years."[35] If DHS were to inspect each facility in order to approve a facility's updated SSP, DHS would need to perform approximately 108 inspections monthly to finish these inspections before receiving updated documents.[36]

The DHS is considering having facilities update their Top-Screen information on this schedule and only submit an updated security vulnerability assessment and site security plan in the case of changes in the Top-Screen information. If so, the number of inspections necessary to approve the updated SSPs would be reduced.

Assessment and Policy Implications

Assessing the future performance of the CFATS program is challenging since it may or may not be similar to the program's past performance. Many external and internal events could dramatically affect program performance. In addition, while the CFATS program has been in place since 2007, significant reforms began in 2012, changing how DHS implemented the program. The data on performance since those reforms are relatively sparse compared with the program's overall duration. Increased efficiencies on the part of the DHS or more effective compliance by regulated facilities could change program performance. Any assessment of the CFATS program is based on the significant assumption that insight about future performance can be informed by analysis of past performance. Even if this assumption is incorrect, however, an analysis based on it may still help to identify potential issues with respect to current agency implementation.

The CFATS program has historically not met DHS-established deadlines for securing high-risk chemical facilities. That said, DHS has now authorized SSPs for almost all facilities receiving a final tier assignment, but authorizing facilities without a final tier assignment will take an unknown amount of time. At the current level of performance, it will likely take approximately two more years to inspect already authorized facilities and approve the SSPs. This conclusion does not consider the impact of compliance inspections, which have only recently begun. Several factors contribute to this conclusion.

Approximately 25% of facilities regulated under CFATS remain in a state of preliminary risk-tier assignment.[37, 38] As described by DHS, this places those facilities at an early stage of review and compliance. Without an effective mechanism for reducing the number of preliminarily tiered facilities, hundreds of high-risk facilities will be in the SSP authorization and approval portion of the regulation rather than in the SSP implementation and compliance portion. The DHS has acknowledged this effect in the short term by changing its target metric from approval of all Tier 1 facility SSPs by the end of FY2013 to approval of over 90% of Tier 1 and Tier 2 facilities with authorized SSPs by the end of FY2014. The DHS appears to have met this new target.

The CFATS system as a whole appears to operate more slowly than would be necessary to process all regulated facilities within the compliance inspection timeframe established in the CFATS regulation. The most obvious factor reducing throughput appears to be the rate of authorization inspection. The DHS continues to authorize SSPs at a rate equal to or greater than the number of authorization inspections conducted. Since the number of SSPs approved is less than the number of authorization inspections, it appears that multiple inspections are necessary to approve some facilities. Therefore, the rate at which authorization inspections need to be conducted limits the rate of overall compliance with the CFATS program.

Analysis of the available data indicates that during some time periods, the number of authorization inspections exceeds the number of SSP approvals, while during other periods, the opposite is true. This suggests that in addition to the time necessary to perform an authorization inspection, time is required for approval of the SSPs. If so, increasing the number of authorization inspections alone may reveal a secondary capacity gap in the rate at which DHS can process SSP approvals.

The fact that SSP authorizations and authorization inspections roughly parallel each other suggests that DHS does not have significant additional capacity in its authorization inspection process. If the authorization inspection process had additional capacity, the number of authorization inspections would not be limited by the number of authorized SSPs and would equal or surpass that number, assuming that some facilities require multiple authorization inspections prior to SSP approval. Increasing authorization inspection capacity might serve to highlight other potential issues within the CFATS process, such as delays in processing information from authorization inspections and issuing letters of approval. Inspection capacity might be increased by employing more inspectors or by process improvements allowing for more inspections done by the same number of inspectors.

Once compliance inspections start in earnest, the lack of additional inspection capacity may lead to further challenges. The same inspector cadre is responsible for both compliance inspections and authorization inspections. Increasing the required number of inspections by beginning compliance inspections could reduce the availability of staff for completing authorization inspections.

Because DHS has set a short interval for Top-Screen resubmission (every two years for Tier 1 and Tier 2 and every three years for Tier 3 and Tier 4) it appears necessary for initial submissions from regulated facilities to be processed within that time frame. If not, a significant backlog of activity seems likely given the current rates of inspection and approval. As stated above, to process all facilities within the time frame of resubmission, DHS would need to perform 108 inspections per month, higher than the current authorization inspection rate, to avoid developing a backlog relative to facility resubmissions. Should DHS not require all facilities to resubmit updated SVAs and SSPs, this may allow DHS to have a lower inspection rate without developing such a backlog.

As DHS makes progress in approving SSPs and as it begins to gather experience in compliance inspections, it may become possible to project future performance with greater confidence.

End Notes

[1] P.L. 109-295, Department of Homeland Security Appropriations Act, 2007, §550.

[2] P.L. 113-254, Protecting and Securing Chemical Facilities from Terrorist Attacks Act of 2014.

[3] In July 2007, DHS testified that formal site inspections of a selected group of facilities would begin by the end of the calendar year (Testimony of Robert B. Stephan, Assistant Secretary for Infrastructure Protection, National Protection and Programs Directorate, Department of Homeland Security, before the House Committee on Homeland Security, Subcommittee on Transportation Security and Infrastructure, July 24, 2007). In December 2007, DHS testified that facility inspection would begin in fall of 2008 (Testimony of Robert B. Stephan, Assistant Secretary for Infrastructure Protection, National Protection and Programs Directorate, Department of Homeland Security, before the House Committee on Homeland Security, Subcommittee on Transportation Security and Infrastructure, December 13, 2007). In 2009, DHS testified that inspections would begin in the first quarter of FY2010 (Testimony of Philip Reitinger, Deputy Under Secretary, National Protection and Programs Directorate, Department of Homeland Security, before the House Committee on Homeland Security, June 16, 2009). The first authorization inspection took place in July 2010 (House Committee on Energy and Commerce, Subcommittee on Environment and the Economy, Evaluating Internal Operation and Implementation of the Chemical Facility Anti-Terrorism Standards [CFATS] Program by the Department of Homeland Security, Serial 112-111, February 3, 2012, p. 65).

[4] In 2010, DHS testified that it expected to inspect all Tier 1 facilities by the end of calendar year 2010 (Oral testimony of Rand Beers, Under Secretary, National Protection and Programs Directorate, Department of Homeland Security, before the Senate Committee on Homeland Security and Governmental Affairs, March 3, 2010). In 2011, DHS testified that it expected to inspect all Tier 1 facilities by the end of calendar year 2011 (Oral testimony of Rand Beers, Under Secretary, National Protection and Programs Directorate, Department of Homeland Security, before the House Committee on Homeland Security, Subcommittee on Cybersecurity, Infrastructure Protection, and Security Technologies, February 11, 2011). In 2013, DHS testified that it planned to have all Tier 1 facilities approved by October 2013 (Testimony of Rand Beers, Under Secretary, and David Wulf, Director, Infrastructure Security Compliance Division, National Protection and Programs Directorate, Department of Homeland Security, before the House Committee on Energy and Commerce, Subcommittee on Environment and the Economy, March 14, 2013). The DHS also planned to have all Tier 1 and Tier 2 facilities approved by May 2014 (Office of the Inspector General, Department of Homeland Security, Effectiveness of the Infrastructure Security Compliance Division's Management Practices to Implement the Chemical Facility Anti-Terrorism Standards Program, OIG-13-55, March 2013, p. 22). The DHS did not meet these milestones. It estimated that, by the end of FY2014, it would have approved over 90% of all Tier 1 and Tier 2 facilities that have authorized site security plans (Communication between Office of Legislative Affairs, Department of Homeland Security, and CRS, October 25, 2013). The DHS met this milestone.

[5] P.L. 109-295, Department of Homeland Security Appropriations Act, 2007, §550.

[6] According to the White House Office of Management and Budget, a performance standard is a standard that states requirements in terms of required results with criteria for verifying compliance but without stating the methods for achieving required results. A performance standard may define the functional requirements for the item, operational requirements, and/or interface and interchangeability characteristics. A performance standard may be

viewed in juxtaposition to a prescriptive standard which may specify design requirements, such as materials to be used, how a requirement is to be achieved, or how an item is to be fabricated or constructed. (Office of Management and Budget, The White House, "Federal Participation in the Development and Use of Voluntary Consensus Standards and in Conformity Assessment Activities," Circular A-119, February 10, 1998). For example, a performance standard might require that a facility perimeter be secured, whereas a prescriptive standard might dictate the height and type of fence to be used to secure the perimeter.

[7] 72 Federal Register 17688–17745 (April 9, 2007).

[8] See, for example, American Chemistry Council, Alternate Security Program (ASP) Guidance for CFATS Covered Chemical Facilities, December 2012.

[9] 72 Federal Register 17688-17745 (April 9, 2007) at 17691. The DHS is considering having facilities update their Top-Screen information on this schedule and only submit an updated security vulnerability assessment and site security plan in the case of changes in the Top-Screen information.

[10] P.L. 113-254, Protecting and Securing Chemical Facilities from Terrorist Attacks Act of 2014.

[11] The DHS Office of Inspector General criticized prior language used by ISCD to describe the CFATS program's progress, calling it "confusing" and "ambiguous." Office of Inspector General, Department of Homeland Security, Effectiveness of the Infrastructure Security Compliance Division's Management Practices to Implement the Chemical Facility Anti-Terrorism Standards Program, OIG-13-55, March 2013.

[12] In 2011, DHS created "conditional authorization," whereby DHS added specific technical conditions that must be addressed during an authorization inspection. The DHS has not completed its rulemaking regarding personnel surety, so site security plans currently receive "conditional authorization."

[13] Testimony of David Wulf, Director, Infrastructure Security Compliance Division, National Protection and Programs Directorate, Department of Homeland Security, before the House Committee on Homeland Security, Subcommittee on Cybersecurity, Infrastructure Protection, and Security Technologies, August 1, 2013.

[14] The DHS received submissions from 36,189 unique facilities as of December 2013. Communication between Office of Legislative Affairs, Department of Homeland Security, and CRS, January 9, 2014.

[15] As of October 20, 2013, DHS had completed more than 600 facility requests for redetermination that resulted in the non-regulation or retiering of many facilities (Office of Infrastructure Protection, National Protection and Programs Directorate, Department of Homeland Security, Chemical Facility Anti-Terrorism Standards (CFATS) and Ammonium Nitrate Security Program Updates, October 2013).

[16] Department of Homeland Security, Chemical Facility Anti-Terrorism Standards, January 2015.

[17] Office of Inspector General, Department of Homeland Security, Effectiveness of the Infrastructure Security Compliance Division's Management Practices to Implement the Chemical Facility Anti-Terrorism Standards Program, OIG-13-55, March 2013, p. 25.

[18] As of December 29, 2014, 89% of Tier 1, 80% of Tier 2, 75% of Tier 3, and 73% of Tier 4 facilities had a final tier assignment (CRS analysis of DHS data provided by Office of Legislative Affairs, Department of Homeland Security, December 29, 2014).

[19] 75 Federal Register 1552-1555 (January 12, 2010).

[20] Communication between Office of Legislative Affairs, Department of Homeland Security, and CRS, January 9, 2014.

[21] This assumes that DHS is presenting output data in terms of net authorizations, where authorized facilities that leave the program reduce the number of authorizations reported. If DHS is presenting process data, where DHS reports the number of authorizations performed that month irrespective of authorized facilities leaving the program, then the number of authorizations will continue to grow. In this latter case, the number of authorizations will surpass the number of facilities, but some facilities will still lack authorized site security plans.

[22] CRS calculation based on number of authorized site security plans reported between February 1, 2014, and February 2, 2015.

[23] CRS calculation based on number of authorized site security plans reported between August 1, 2014, and February 2, 2015.

[24] CRS calculation based on number of authorized site security plans reported between January 5, 2015, and February 2, 2015.

[25] House Committee on Energy and Commerce, Subcommittee on Environment and the Economy, The Chemical Facility Anti-Terrorism Standards Program: A Progress Report, Serial No. 112-172, September 11, 2012, p. 112.

[26] This assumes that DHS is presenting output data in terms of net approvals, where approved facilities that leave the program reduce the number of approvals reported. If DHS is presenting process data, where DHS reports the number of approvals performed that month irrespective of approved facilities leaving the program, then the number of approvals will continue to grow. In this latter case, the number of approvals will surpass the number of facilities, but some facilities will still lack approved site security plans.

[27] CRS calculation based on number of approved site security plans reported between February 1, 2014, and February 2, 2015.

[28] CRS calculation based on number of approved site security plans reported between August 1, 2014, and February 2, 2015.

[29] CRS calculation based on number of approved site security plans reported between January 5, 2015, and February 2, 2015.

[30] Based on an approval rate of 30 to 40 SSPs monthly. Government Accountability Office, Critical Infrastructure Protection: DHS Efforts to Assess Chemical Security Risk and Gather Feedback on Facility Outreach Can Be Strengthened, GAO-13-353, April 2013.

[31] Testimony of Suzanne Spaulding, Under Secretary for National Protection and Programs, and David Wulf, Director, Infrastructure Security Compliance Division, Office of Infrastructure Protection, National Protection and Programs Directorate, Department of Homeland Security before the Senate Committee on Homeland Security and Governmental Affairs, May 14, 2014.

[32] Communication between Office of Legislative Affairs, Department of Homeland Security, and CRS, December 15, 2014.

[33] Government Accountability Office, Critical Infrastructure Protection: DHS Efforts to Assess Chemical Security Risk and Gather Feedback on Facility Outreach Can Be Strengthened, GAO-13-353, April 2013.

[34] The DHS had approved the SSPs of 362 CFATS facilities as of November 19, 2013.

[35] 72 Federal Register 17688-17745 (April 9, 2007) at 17691.

[36] Based on 494 preliminary and final Tier 1 and Tier 2 facilities and 3,140 preliminary and final Tier 3 and Tier 4 facilities (Communication between Office of Legislative Affairs, Department of Homeland Security, and CRS, December 29, 2014).

[37] CRS analysis based on DHS data of December 2014 (Communication between Office of Legislative Affairs, Department of Homeland Security, and CRS, December 29, 2014).

[38] Approximately 405 of these are facilities with aboveground gasoline storage tanks. The DHS has identified these facilities as preliminarily high risk but has indefinitely extended their compliance deadlines (75 Federal Register 1552- 1555 (January 12, 2010)).

In: Chemical Facility Anti-Terrorism Standards ISBN: 978-1-63484-465-9
Editor: Joanne Nelson

Chapter 2

CRITICAL INFRASTRUCTURE PROTECTION: DHS ACTION NEEDED TO VERIFY SOME CHEMICAL FACILITY INFORMATION AND MANAGE COMPLIANCE PROCESS*

United States Government Accountability Office

WHY GAO DID THIS STUDY

Thousands of facilities have hazardous chemicals that could be targeted or used to inflict mass casualties or harm surrounding populations in the United States. DHS established the CFATS program to, among other things, identify and assess the security risk posed by chemical facilities. Within DHS, ISCD oversees this program.

GAO was asked to assess the CFATS program. This report addresses, among other things, the extent to which DHS has (1) categorized facilities as subject to the CFATS regulation, and (2) approved site security plans and conducted compliance inspections. GAO reviewed laws, regulations, and program documents; randomly selected data submitted to ISCD by facilities from 2007 to 2015, tested the data's reliability; and generated estimates for the entire population of facilities, and interviewed officials responsible for

---

* This is an edited, reformatted and augmented version of a United States Government Accountability Office publication, No. GAO-15-614, dated July 2015.

overseeing, identifying, categorizing, and inspecting chemical facilities from DHS headquarters and in California, Maryland, Oregon, and Texas (selected based on geographic location and other factors).

What GAO Recommends

GAO recommends, among other things, that DHS (1) verify the Distance of Concern reported by facilities is accurate and (2) document processes and procedures for managing compliance with site security plans. DHS concurred with GAO's recommendations and outlined steps to address them.

What GAO Found

Since 2007, the Office of Infrastructure Protection's Infrastructure Security Compliance Division (ISCD), within the Department of Homeland Security (DHS), has identified and collected data from approximately 37,000 chemical facilities under its Chemical Facility Anti-Terrorism Standards (CFATS) program and categorized approximately 2,900 as high-risk based on the collected data. However, ISCD used unverified and self-reported data to categorize the risk level for facilities evaluated for a toxic release threat. A toxic release threat exists where chemicals, if released, could harm surrounding populations. One key input for determining a facility's toxic release threat is the Distance of Concern (distance) that facilities report—an area in which exposure to a toxic chemical cloud could cause serious injury or fatalities from short-term exposure. ISCD requires facilities to calculate the distance using a web-based tool and following DHS guidance. ISCD does not verify facility-reported data for facilities it does not categorize as high-risk for a toxic release threat. However, following DHS guidance and using a generalizable sample of facility-reported data in a DHS database, GAO estimated that more than 2,700 facilities (44 percent) of an estimated 6,400 facilities with a toxic release threat misreported the distance. By verifying that the data ISCD used in its risk assessment are accurate, ISCD could better ensure it has identified the nation's high-risk chemical facilities.

ISCD has made substantial progress approving site security plans but does not have documented processes and procedures for managing facilities that are noncompliant with their approved site security plans. Site security plans outline, among other things, the planned measures that facilities agree to implement to address security vulnerabilities. As of April 2015, GAO estimates that it could take between 9 and 12 months for ISCD to review and approve security plans for approximately 900 remaining facilities—a substantial improvement over the previous estimate of 7 to 9 years GAO reported in April 2013. ISCD officials attributed the increased approval rate to efficiencies in ISCD's security plan review process, updated guidance, and a new case management system. Further, ISCD began conducting compliance inspections in September 2013, but does not have documented processes and procedures for managing the compliance of facilities that have not implemented planned measures outlined in their site security plans. According to the nature of violations thus far, ISCD has addressed noncompliance on a case-by-case basis. Almost half (34 of 69) of facilities ISCD inspected as of February 2015 had not implemented one or more planned measures by deadlines specified in their approved site security plans and therefore were not fully compliant with their plans. GAO found variations in how ISCD addressed these 34 facilities, such as how much additional time the facilities had to come into compliance and whether or not a follow-on inspection was scheduled. Such variations may or may not be appropriate given ISCD's case-by-case approach, but having documented processes and procedures would ensure that ISCD has guidelines by which to manage noncompliant facilities and ensure they close security gaps in a timely manner. Additionally, given that ISCD will need to inspect about 2,900 facilities in the future, having documented processes and procedures could provide ISCD more reasonable assurance that facilities implement planned measures and address security gaps.

ABBREVIATIONS

ASP	Alternative Security Program
CFATS	Chemical Facility Anti-Terrorism Standards
CI	Compliance Inspection
COI	Chemical of Interest
CSAT	Chemical Security Assessment Tool
CVI	Chemical-terrorism Vulnerability Information
DHS	Department of Homeland Security
EPA	Environmental Protection Agency
GPRA	Government Performance and Results Act

IP	Office of Infrastructure Protection
ISCD	Infrastructure Security Compliance Division
NIPP	National Infrastructure Protection Plan
NPPD	National Protection and Programs Directorate
RBPS	Risk-Based Performance Standards
SSP	Site Security Plan

* * *

July 22, 2015
Congressional Requesters

Facilities that produce, use, or store hazardous chemicals could be of particular interest to terrorists who are intent on using toxic chemicals to inflict mass casualties in the United States. These chemicals could be released from a facility to cause harm to surrounding populations, stolen and used as chemical weapons or the ingredients to make chemical weapons, or stolen and used to build an improvised explosive device. For example, on April 19, 1995, a domestic terrorist used ammonium nitrate fertilizer mixed with fuel oil to blow up a federal building in Oklahoma City, Oklahoma. The explosion killed 168 people and injured hundreds more. The Chemical Facility Anti-Terrorism Standards (CFATS) program, established pursuant to the Department of Homeland Security (DHS) Appropriations Act, 2007, enables DHS to, among other things, identify chemical facilities and assess the security risk posed by each, categorize the facilities into risk-based tiers, and inspect the high-risk facilities to ensure compliance with regulatory requirements.[1] The Protecting and Securing Chemical Facilities from Terrorist Attacks Act of 2014, enacted in December 2014, in effect, reauthorized the CFATS program for an additional 4 years while also imposing additional implementation requirements on DHS for the program.[2] The Office of Infrastructure Protection (IP), through its Infrastructure Security Compliance Division (ISCD), within DHS's National Protection and Programs Directorate (NPPD), oversees the CFATS program.

In July 2012, we reported that ISCD had efforts under way to address challenges in implementing the CFATS program—such as delays approving site security plans—that were highlighted in a leaked internal ISCD memorandum. 3 We found that ISCD was in the early stages of implementing actions to address challenges and had not yet established performance measures to assess results. We recommended that ISCD explore opportunities to develop performance measures and assess results, where practical. ISCD agreed with our recommendation and, in response, developed an operating plan that included information on how ISCD planned to measure performance. Subsequently, in April 2013, we reported that ISCD had categorized about 3,500 chemical facilities as high-risk, but that the approach ISCD used did not consider all elements of risk (consequence, threat, and vulnerability),[4] as called for by the *National Infrastructure Protection Plan* (NIPP).[5] We recommended, among other things, that ISCD enhance its risk assessment approach to incorporate all elements of risk, and conduct a peer review of the program to validate and verify ISCD's risk assessment approach. ISCD agreed with our recommendations and, as of May 2015, was taking steps intended to address the recommendations.

The Protecting and Securing Chemical Facilities from Terrorist Attacks Act includes a provision for GAO to assess implementation of the act and amendments made by the act and submit an initial report to Congress not later than 180 days after enactment. 6 In addition, the explanatory statement accompanying the Consolidated and Further Continuing Appropriations Act, 2013 (Public Law 113-6), provides that GAO is to continue its ongoing effort to examine the extent to which DHS has made progress and encountered challenges in developing a viable CFATS program.[7] We performed this review in response to both provisions, as well as additional requests for reviews related to chemical security and ammonium nitrate. This report assesses the extent to which DHS has

- identified and categorized facilities as subject to the CFATS regulation, and
- approved site security plans, conducted compliance inspections, and measured results.

Appendix I of this report provides an update of ISCD actions under way to address the Protecting and Securing Chemical Facilities from Terrorist Attacks Act. To address our first objective, we reviewed laws and regulations applicable to facilities that possess or store chemicals, as well as Executive Order 13650, to determine how ISCD is to identify and categorize chemical facilities.[8] To evaluate how ISCD has identified facilities that could potentially be subject to CFATS, we reviewed and analyzed ISCD documentation related to its efforts to identify facilities, including documentation on coordinating with other federal agencies that regulate chemical facilities, and the Executive Order 13650 *Report for the President: Actions to Improve Chemical Facility*

Safety and Security—a Shared Commitment.[9] To evaluate how ISCD has categorized chemical facilities, we reviewed and analyzed documentation ISCD provides to facilities intended to guide them in submitting data that ISCD uses to preliminarily categorize facilities as high-risk, which renders them subject to additional requirements under the CFATS regulation during its preliminary screening process.[10] As part of this effort, we obtained data submitted by facilities to ISCD in accordance with CFATS. We analyzed a statistical sample of these data to determine the extent to which ISCD used facility-submitted data to make categorization decisions that are reproducible, defensible, and free from significant errors and omissions, as called for by the *National Infrastructure Protection Plan.* We also compared ISCD's facility categorization efforts to project management guidance.[11] We assessed the reliability of ISCD data by reviewing relevant documentation and interviewing knowledgeable officials. We concluded that these data were sufficiently reliable for the purposes of this report. We also interviewed ISCD officials to confirm our understanding of the documents and data provided, and actions ISCD has taken to identify and categorize chemical facilities.

To address our second objective, we reviewed laws and regulations applicable to how DHS is to approve site security plans and ensure compliance with the CFATS regulation. We analyzed ISCD data to identify the number of site security plans approved per month, and used the results of our analysis to estimate the number of months it could take ISCD to approve remaining site security plans. We note limitations in our analysis, including that our estimate does not take into consideration resource constraints ISCD may face as it implements the Ammonium Nitrate Security Program once the rule establishing the program has been issued.[12] We analyzed compliance inspection reports to identify the extent to which the reports identify non-compliance among facilities that have undergone compliance inspections as part of the CFATS program. We compared ISCD's processes and procedures for tracking and monitoring noncompliant facilities to the NIPP, which calls for risk assessment approaches to be documented, reproducible, and defensible, and to standards in *Standards for Internal Control in the Federal Government* for documenting transactions and significant decisions.[13] To determine how DHS has measured results, we compared ISCD's performance measure with criteria in the NIPP for evaluating the effectiveness of risk management efforts by, among other things, collecting performance data to assess progress in achieving outputs and outcomes. We interviewed ISCD headquarters officials responsible for overseeing the CFATS program to confirm our understanding of their processes and systems for authorizing, approving, and inspecting facilities for compliance. Finally, we conducted four site visits to observe two authorization inspections and two compliance inspections. We selected inspection locations in California, Oregon, Maryland, and Texas based on geographic dispersion and to cover different types of chemical facilities regulated by the CFATS program. While the information obtained from these inspections cannot be generalized to all inspections, it provides insight and context on how ISCD conducts CFATS inspections. For further details on our scope and methodology, see appendix II.

We conducted this performance audit from September 2014 to July 2015 in accordance with generally accepted government auditing standards. Those standards require that we plan and perform the audit to obtain sufficient, appropriate evidence to provide a reasonable basis for our findings and conclusions based on our audit objectives. We believe that the evidence obtained provides a reasonable basis for our findings and conclusions based on our audit objectives.

BACKGROUND

The CFATS program is intended to ensure the security of the nation's chemical infrastructure by identifying, assessing the risk posed by, and requiring the implementation of measures to protect high-risk chemical facilities. Section 550 of the DHS Appropriations Act, 2007, required DHS to issue regulations establishing risk-based performance standards for chemical facilities that, as determined by DHS, present high levels of risk, and required vulnerability assessments and the development and implementation of site security plans for such facilities.[14] DHS published the CFATS interim final rule in April 2007 and appendix A to the rule, published in November 2007, lists 322 chemicals of interest (COI) and the screening threshold quantities amount for each.[15] According to DHS and subject to certain statutory exclusions, all facilities that manufacture chemicals of interest as well as facilities that store or use such chemicals as part of their daily operations may be subject to CFATS.[16] In general, however, only chemical facilities determined to possess a requisite quantity of COI (that is, the screening threshold quantity) and that are subsequently determined to present high levels of security risk—that is, covered facilities—are subject to the more substantive requirements of the CFATS regulation.[17] The Protecting and Securing Chemical Facilities from Terrorist Attacks Act of 2014, enacted in December 2014, amended the Homeland Security Act of 2002 by adding the Chemical

Facility Anti-Terrorism Standards as Title XXI and, in effect, authorizing the program for an additional 4 years.[18] Among other things, the Act expressly repeals DHS's authority to implement the program under section 550 of the DHS Appropriations Act, 2007, but also expressly provides that the CFATS regulation promulgated under that authority shall remain in effect unless otherwise amended, consolidated, or repealed. Consequently, while the Act imposes new and additional responsibilities on DHS to implement the CFATS program, the program continues to be implemented by ISCD under the existing regulatory framework.[19]

In addition to implementing the CFATS program, ISCD is also to manage the Ammonium Nitrate Security Program—a program DHS is to establish pursuant to section 563 of the DHS Appropriations Act, 2008.[20] While the CFATS program covers facilities with certain thresholds of ammonium nitrate, the Ammonium Nitrate Security Program is to regulate the sale and transfer of ammonium nitrate by an ammonium nitrate facility for the purpose of preventing the misappropriation or use of ammonium nitrate in an act of terrorism.[21] Among other things, the statute authorizing the ammonium nitrate program authorizes DHS to require individuals who purchase, sell, or transfer ammonium nitrate to register with DHS and submit to vetting against the Terrorist Screening Database, and requires the owners of ammonium nitrate facilities to maintain records on each sale or transfer of ammonium nitrate and to report any identified theft or loss of ammonium nitrate to appropriate federal authorities.[22] DHS issued a proposed rule for the Ammonium Nitrate Security Program in August 2011 and, as of July 2015, has yet to establish the program.[23]

The CFATS Regulation and Process

The CFATS regulation outlines how ISCD is to administer the CFATS program. Specifically, any facility that possess any of the 322 chemicals of interest (COI) in quantities that meet or exceed the screening threshold quantities established by DHS for those COI are required to use ISCD's Chemical Security Assessment Tool (CSAT)—a web-based application through which owners and operators of facilities with COI are to provide information about the facility—to complete a Top-Screen.[24] The Top-Screen is the initial screening tool whereby a chemical facility in possession of a COI at the requisite thresholds is to provide ISCD data, including the name and location of the facility and the chemicals and their quantities at the site.

ISCD's risk assessment approach, which relies on data from the Top-Screen, among other sources, is based on three security issues: (1) release (toxic, flammable, and explosive) chemicals with the potential for impacts within and beyond a facility; (2) theft or diversion; and (3) sabotage, depending on the type of risk associated with the COI.

- **Release:** For the release threat, ISCD's approach assumes that a terrorist will release the COI at the facility and then estimates the risk to the facility and surrounding population. Facilities with toxic release chemicals are to calculate and report in their Top-Screen submission the Distance of Concern—which represents the radius of an area in which exposure to a toxic chemical cloud from a release event could cause serious injury or fatalities from short-term exposure. ISCD uses the Distance of Concern to estimate the number of fatalities from an intentional toxic release and to categorize the risk posed by this facility.[25] The Top-Screen directs respondents to use an online tool called RMP*Comp to calculate the Distance of Concern.[26] RMP*Comp takes inputs such as the quantity of chemical that could be released and the surrounding terrain type to determine the Distance of Concern.
- **Theft or diversion:** For theft or diversion, the approach assumes that a terrorist will steal or have the COI diverted to him or herself and then estimates the risk of a terrorist attack using the COI to cause the most harm at an unspecified off-site location.
- **Sabotage:** For sabotage, the approach assumes that a terrorist will cause water to be mixed with a COI that is shipped from the facility, creating a toxic release at an unspecified location, and then estimates the risk to a medium-sized U.S. city.

If, according to ISCD's automated assessment of information provided via the Top-Screen, the facility is preliminarily categorized to be high-risk it becomes a "covered chemical facility," and ISCD is to notify the facility of its preliminary placement in one of four risk-based tiers—tier 1, 2, 3, or 4. If ISCD does not categorize the chemical facility as high-risk, ISCD does not assign the facility to one of these four risk-based tiers and the facility is not subject to additional requirements under the CFATS regulation.[27] Facilities that ISCD preliminarily categorizes to be high-risk—covered chemical facilities—are required to then complete the CSAT security vulnerability assessment, which includes the identification of potential critical assets at the facility, and a related vulnerability analysis.[28] ISCD is to review the security vulnerability assessment to confirm and notify the facility as to whether the facility remains categorized as high-risk and, if so, about its final placement in one of the four tiers.

Once a covered chemical facility is assigned a final tier, the facility may use CSAT to submit a site security plan (SSP) or submit an Alternative Security Program in lieu of the CSAT SSP.[29] The security plan is to describe the existing and planned security measures to be implemented to address the vulnerabilities identified in the security vulnerability assessment, and identify and describe how existing and planned security measures selected by the facility are to address the applicable risk-based performance standards.[30] To meet risk-based performance standards, covered facilities may choose the security programs or processes they deem appropriate to address the performance standards so long as ISCD determines that the facilities achieve the requisite level of performance on each of the applicable areas in their existing and agreed-upon planned measures.

To determine whether facilities achieve the requisite level of performance for each of the applicable areas, ISCD is to conduct a preliminary review of the facility's security plan to determine whether it meets the risk-based regulatory requirements. If these requirements appear to be satisfied, ISCD is to issue a letter of authorization for the plan, and conduct an authorization inspection of the facility to determine whether to approve the plan. Upon inspection of the facility, if ISCD determines that the plan satisfies the CFATS requirements, it will issue a letter of approval to the facility, which is to then implement the approved SSP. If ISCD determines that the plan does not satisfy CFATS requirements, ISCD then notifies the facility of any deficiencies and the facility must submit a revised plan for correcting them.[31]

Following ISCDs approval of a facility's SSP, in order to assess compliance with CFATS requirements as addressed through the approved SSP, ISCD conducts a compliance inspection (CI) of the covered facilities.[32] CIs are to follow specific standard operating procedures that focus on verifying existing measures and the implementation and effectiveness of planned measures, including dates implemented, as well as verifying and reviewing any significant changes in the facility's security posture. If through a compliance inspection it is determined a facility has not fully implemented security measures as outlined in its approved site security plan, ISCD is to provide the facility with written notification that clearly identifies the deficiencies in the SSP and will work with the facility towards achieving full compliance or, if warranted, take enforcement action.[33] For example, the

CFATS regulation provides that an order compelling a facility to take appropriate action may be issued if the facility was found to be in violation of any part of the regulation. If a facility were to violate this initial order, an order assessing a civil penalty of up to $25,000 per day or to cease operations, or both, may be issued.[34] According to ISCD guidance, inspectors conducting a CI are to document in a CI report any necessary enforcement actions that may result from the inspection, including, among others, security measures not implemented in accordance with the SSP. ISCD guidance further provides that inspectors are not to recommend enforcement actions where issues identified during a CI are quickly remedied on-site; however, inspectors are to document their finding in the CI report. Inspection teams are to submit their CI report to ISCD management for review and approval within 25 business days from the inspection completion. ISCD management is to then determine whether a facility is in compliance with their approved SSP or whether to take enforcement actions. Figure 1 illustrates the CFATS regulatory process.

ISCD Has Taken Steps to Identify Chemical Facilities but Used Unverified Data to Categorize Thousands of Facilities

ISCD Has Identified about 37,000 Chemical Facilities, but Plans Additional Identification Efforts

Since 2007, when ISCD began identifying chemical facilities to determine which facilities present a high risk and therefore should be subject to further regulation under CFATS, about 37,000 facilities have submitted a Top-Screen but ISCD officials acknowledged some facilities may have failed to do so. According to these officials, as of April 2015, ISCD had received most Top-Screens within the first 3 years of the program; specifically, ISCD received about 88 percent of Top-Screens from 2007 through 2009.[35] ISCD officials told us they believe the 37,000 facilities that have submitted Top-Screens represent most facilities subject to CFATS. However, as we previously reported, ISCD officials have acknowledged some facilities may not comply with the requirement to submit a Top-Screen and therefore some facilities may not be included in ISCD's data, but the magnitude of potential Top-Screen noncompliance is not known.[36]

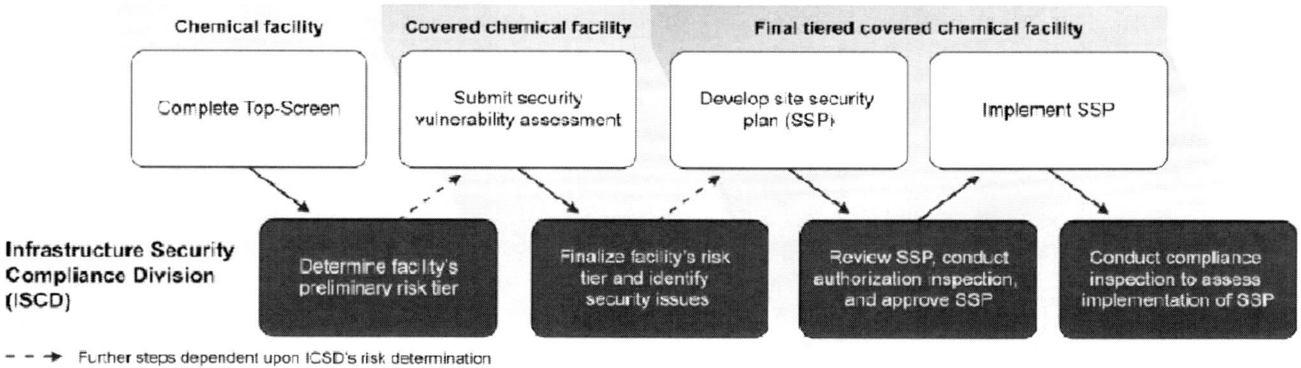

Source: GAO analysis of CFATS regulatory process. | GAO-15-614.

Figure 1. Department of Homeland Security's (DHS) Chemical Facility Anti-Terrorism Standards (CFATS) Process.

In response to Executive Order 13650, in May 2014, ISCD outlined a number of efforts it had taken and planned to take to identify potentially noncompliant chemical facilities. These efforts included comparing facilities data from relevant federal partners and from state entities against ISCD's current database of chemical facilities to identify potentially noncompliant chemical facilities, among others.[37] ISCD officials told us ISCD consulted with the Occupational Safety and Health Administration; the Department of Transportation; the Department of Agriculture; the Bureau of Alcohol, Tobacco, Firearms and Explosives; and the Environmental Protection Agency (EPA) to determine the extent to which ISCD could identify all chemical facilities using data held by those agencies. According to ISCD officials, ISCD also reached out to all 50 states' Homeland Security Advisors to request lists of chemical facilities maintained by the states.

According to ISCD officials, as of April 2015, ISCD completed some of these efforts, such as conducting cross-agency compliance analysis with EPA and analyzing facility lists provided by New York, New Jersey, and Texas. Although these efforts resulted in the identification of chemical facilities subject to CFATS that had not submitted a Top-Screen, ISCD officials told us the work resulted in few new covered facilities. For example, in 2013, after conducting cross-agency compliance analysis with EPA, ISCD identified and notified about 3,300 chemical facilities as potentially noncompliant with the requirement to submit a Top-Screen. As shown in figure 2, in response to notification by ISCD 1,571 facilities reported they had previously submitted a Top-Screen and another 361 claimed they are not currently subject to the requirement to submit a TopScreen.[38] An additional 374 facilities have not submitted a Top-Screen for other reasons such as the facility closed or the facility does not have chemicals of interest above the screening threshold quantity. Of the 1,056 facilities required to submit a Top-Screen as a result of ISCD's notification, representing approximately 30 percent of all facilities notified, ISCD categorized 2.3 percent, 24 facilities, as high-risk.[39]

ISCD officials told us that although the effort to compare EPA and CFATS facility data did result in more than 1,000 chemical facilities submitting Top-Screens, they noted that efforts to identify chemical facilities have resulted in few new covered facilities. Because of this, ISCD officials are reevaluating how ISCD conducts federal-and state-level cross-agency compliance analysis as a method for identifying chemical facilities. While they plan to continue these efforts and the California Environmental Protection Agency recently provided ISCD with information on over 46,000 facilities regulated by the state, ISCD is currently assessing which approaches for identifying potentially noncompliant facilities will result in a high return on investment. The officials cited two challenges to cross-agency matching. First, while other federal agencies collect some information on chemical facilities, differing program goals and data formats may limit the utility of these data in identifying chemical facilities for the CFATS program. Second, according to ISCD officials, 14 of 50 states responded to ISCD's request for facility lists.[40] To continue identifying chemical facilities for which the submission of a Top-Screen is required, ISCD plans to designate a lead staff member to oversee efforts to identify such facilities, measure effectiveness of efforts relative to the costs and benefits, and recommend a long-term future strategy. ISCD officials told us they also have hired a third-party vendor to pilot an effort to analyze supply-chain data to identify noncompliant facilities.

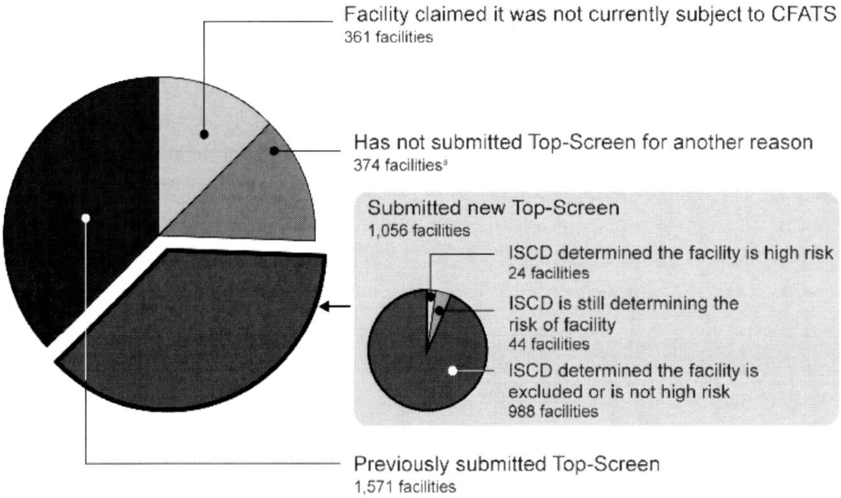

Source: GAO analysis of Infrastructure Security Compliance Division (ISCD) data. | GAO-15-614.

[a] Reasons why these facilities did not submit Top-Screens varied. ISCD stated that data for 244 facilities was no longer accurate because, for example, the facility had been sold or closed. Another 67 facilities indicated their intent to file a Top-Screen; 61 facilities did not respond to ISCD's notification; 1 facility requested ISCD assistance in complying with CFATS requirements, and notification delivery failed for another facility.

Figure 2. Result of Cross-Agency Effort to Identify Chemical Facilities Subject to Regulation under the Chemical Facilities Security Regulations (CFATS) Program.

ISCD Used Self-Reported Data That Is Not Verified to Categorize Thousands of Facilities

ISCD has used self-reported and unverified data to determine the risk for facilities with a toxic release threat. As described earlier, approximately 37,000 facilities have submitted Top-Screens to ISCD. Of these facilities, we estimate that more than 6,400 facilities hold a chemical at or above the screening threshold quantity that could pose a toxic chemical release threat.[41] As part of the Top-Screen, ISCD requires these facilities to self-report the Distance of Concern, which represents the radius of an area in which exposure to a toxic chemical cloud from a release event could cause serious injury or fatalities from short-term exposure.[42] As part of its risk assessment of facilities with toxic release chemicals, ISCD uses the Distance of Concern to determine the consequences from an intentional toxic release. Using DHS guidance and Top-Screen information stored in a DHS database—which contains facility-reported information such as chemicals and quantities in their possession, and the Distance of Concern—we recalculated the Distance of Concern for a generalizable sample of facilities and compared these results to what facilities reported. On the basis of these results, we estimate more than 2,700 facilities, or at least 44 percent of facilities with a toxic release chemical, misreported the Distance of Concern.[43] We further estimate that at least 1,200 of the 2,700 misreporting facilities, or about 43 percent, underestimated the Distance of Concern.[44] Figure 3 depicts an example of 1 facility with more than 200,000 pounds of anhydrous ammonia that reported a Distance of Concern of 0.9 miles compared to a minimum possibly correct Distance of Concern of 2.4 miles.

ISCD officials acknowledged that some facilities may have erred when reporting the Distance of Concern. However, ISCD officials told us that they do not verify the Distance of Concern on the Top-Screen because they believe that most facilities with toxic release chemicals above the screening threshold quantities would be familiar with RMP*Comp. According to ISCD officials, chemical facilities use the RMP*Comp for purposes of meeting EPA regulatory requirements. However, guidance published on the DHS web-site directs facilities to calculate the Distance of Concern in a manner that is different from EPA reporting requirements. For example, to satisfy EPA reporting requirements, facilities can include in their Distance of Concern calculation information about passive mitigation systems (such as dikes, enclosures, berms, and drains) that may reduce the output RMP*Comp generates. Conversely, according to DHS guidance, because the Top-Screen is evaluating an intentional release rather than an accidental release, facilities are not to include passive mitigation systems to calculate the Distance of Concern. ISCD officials stated that because RMP*Comp is outside the domain of ISCD oversight, they cannot say with full certainty that the information facilities use to generate the Distance of Concern is accurate. Nevertheless, ISCD has retained all Top-Screen data, including the necessary information that can be used to verify the accuracy of Distances of Concern provided by relevant facilities. ISCD officials also stated that they plan to develop a new model that will incorporate a Distance of Concern calculation into the next iteration of the Top-Screen, originally scheduled to be operational by September 2015, which will not require facilities to calculate and provide a Distance of Concern using RMP*Comp. Instead, the new Top-Screen will calculate the Distance of Concern using chemical information, such as the quantity of chemicals, reported by facilities. However, officials stated ISCD has not yet decided whether to re-screen facilities that hold toxic release chemicals that have already submitted a Top-Screen. As of May 2015, ISCD officials stated that implementation of the new Top-Screen is delayed, and ISCD officials did not have a timeline for implementation.

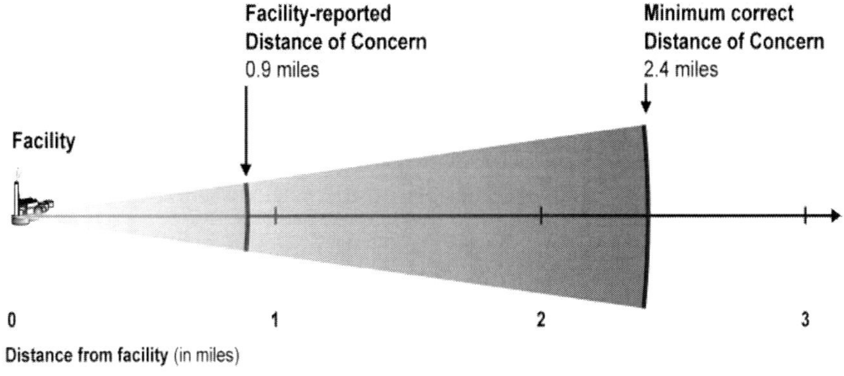

Source: GAO analysis of Infrastructure Security Compliance Division (ISCD) data. | GAO-15-614.

Figure 3. Example of Facility-Reported 0.9-Mile Distance of Concern and Corrected 2.4- Mile Distance of Concern Generated by GAO Using Department of Homeland Security Guidance.

As described earlier, the Distance of Concern is one key input in the risk assessment approach that ISCD uses to preliminarily categorize facilities that could pose a toxic chemical release threat. Facilities ISCD categorizes as high-risk are covered facilities subject to additional requirements under the CFATS regulation, while facilities not categorized as high-risk are, in general, not subject to additional requirements under the CFATS program. According to ISCD officials, ISCD only verifies information in the Top-Screens reported by high-risk facilities, including the Distance of Concern. ISCD officials also told us that because ISCD takes other factors into account, errors in the Distance of Concern may not necessarily result in changes in ISCD's risk-assessment of facilities that misreport the Distance of Concern. According to ISCD, within our sample 1 high-risk facility could be miscategorized based upon an erroneous Distance of Concern. [45] On the basis of ISCD's finding of 1 potentially miscategorized facility within our sample, we estimate that ISCD could have miscategorized 85 high-risk facilities, but potentially up to 543 high-risk facilities that have previously submitted Top-Screens.[46] Additionally, because implementation of the new Top-Screen is delayed, ISCD cannot provide reasonable assurance it will categorize facilities with release toxic chemicals using reliable Distance of Concern data as these facilities submit new Top-Screens.

Standard project management practices include activities such as developing a schedule with milestone dates to identify points throughout the project to reassess efforts under way to determine whether project changes are necessary. Practices such as developing a new target schedule with a forecasted finish date would provide ISCD a more realistic plan to guide the implementation of the new Top-Screen.[47] Additionally, the NIPP risk management framework calls for risk assessment approaches—such as ISCD's risk-based implementation of CFATS—to be reproducible and defensible. Specifically, the NIPP states the risk assessment methodology must produce comparable, repeatable results free from significant errors or omissions. In the interim, before ISCD implements the new Top-Screen and determines whether to rescreen facilities that have already submitted a Top-Screen, identifying potentially miscategorized facilities that may pose a significant security risk and verifying the Distance of Concern these facilities report is accurate could help ensure that ISCD has accurately categorized facilities with the potential to cause the greatest harm. Moreover, ISCD could provide more reasonable assurance it has identified the nation's high-risk chemical facilities and, subsequently, that these facilities take actions to address potential terrorist threats.

ISCD Has Made Substantial Progress Approving Site Security Plans and Conducting Compliance Inspections but Does Not Consistently Ensure Compliance and Effectively Measure Program Results

ISCD Has Substantially Reduced the Time Needed to Approve Remaining Site Security Plans

ISCD has taken actions to improve its processes for reviewing and approving site security plans, which have reduced the amount of time needed to resolve the backlog of unapproved plans. On the basis of ISCD's pace of site security plan approvals in calendar year 2014— between 80 and 100 plans per month[48]—as of April 2015, we estimate that it could take between 9 and 12 months for ISCD to review and approve site security plans for the 929 facilitie currently awaiting approval.[49] This represents a substantial improvement over our previous estimates, in April 2013, when, we reported that, based upon ISCD's estimated approval rate of between 30 and 40 security plans per month, it could take between 7 and 9 years for ISCD to complete reviews of the approximately 3,120 plans in its queue at that time.[50] Figure 4 shows our revised estimate for the time needed to approve plans for unapproved final-tiered facilities as of April 2015—assuming approval rates of 80, 90, and 100 plans per month—as compared to our original April 2013 estimates.

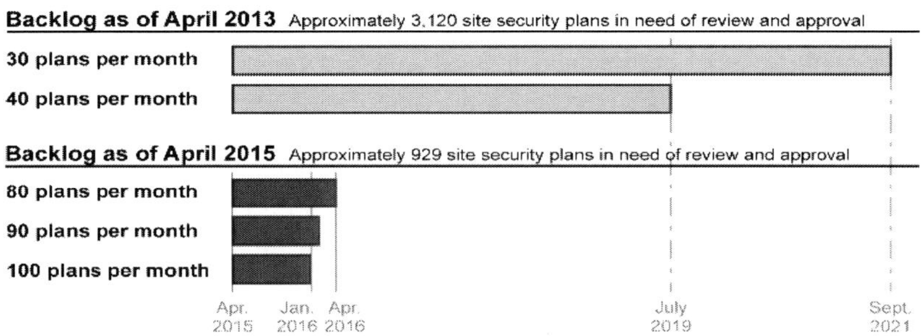

Source: GAO analysis of Infrastructure Security Compliance Division (ISCD) data. | GAO-15-614.

Figure 4. Estimates of Time Needed to Approve Remaining Chemical Facility Anti-Terrorism Standards (CFATS) Site Security Plans.

ISCD officials attributed the increased rate of approvals and reduction in the backlog of facilities awaiting approval to a number of improvements in ISCD's processes. These improvements included new steps ISCD has taken since our last report and planned actions we previously reported in April 2013 and May 2014, but could not assess at that time because they had not yet been fully implemented.[51] For example, ISCD has

- continued the revised site security plan review process implemented in July 2012 in which teams of ISCD headquarters officials review plans by assessing how layers of security measures meet the intent of each of the performance standards;
- issued updates to the online Chemical Security Assessment Tool (CSAT) beginning in March 2014 to make the system more user-friendly and improve facility data collection;
- distributed updated internal guidance and lessons learned on plan approvals to inspectors and plan reviewers;
- transitioned to a new internal case management system in December 2013 that provides improved program and facility management capabilities;
- begun using inspectors alongside ISCD headquarters officials to review site security plans in order to leverage inspectors' knowledge of facility security and role conducting CFATS inspections;
- implemented changes to inspection processes, such as employing smaller inspection teams, conducting preinspection phone calls with facilities to help them prepare for inspections, and enabling inspectors to help facility personnel edit their site security plans during inspections;
- distributed updated guidance to facilities to help them improve their site security plans and worked to expand the use of alternative security programs; and
- worked with corporations that have multiple covered chemical facilities to leverage inspection documents and security procedures that are standard across corporations to expedite the inspection process.

While ISCD has taken actions to reduce the backlog of site security plans, our estimate that ISCD could complete its approval of all current site security plans within 9 to 12 months does not take into account the potential impact of other tasks central to the CFATS program and additional compliance activities for which ISCD is responsible. Specifically, our estimate does not include the time required to:

- identify, categorize, review, and approve site security plans for facilities that have not yet submitted Top-Screens or those that ISCD previously did not categorize as high risk but may now qualify as high-risk in light of the aforementioned errors in the Top-Screen submissions we identified related to toxic release Distance of Concern calculations, and
- review approved site security plans to resolve issues relating to one requirement of the personnel surety performance standard, under which covered facilities are to perform background checks and ensure appropriate credentials for personnel and visitors at their facilities. As part of the personnel surety standard, DHS plans to check for terrorist ties by comparing certain employee information against the Terrorist Screening Database. ISCD currently has measures in place for other screening requirements under the personnel surety standard, and as of May 2015, ISCD was determining how the terrorist screening requirement will be implemented.[52]

In addition, our estimate assumes that the pace of site security plan approvals will not be affected by ISCD's implementation of the Ammonium Nitrate Security Program. As described earlier, ISCD released a proposed rule on August 3, 2011, for how it plans to implement the program. According to ISCD officials, the final rule was initially scheduled to be released in April 2015; as of May 2015, the rule had not been released but, according to the DHS semiannual regulatory agenda published in June 2015, a final rule for the program is anticipated in October 2015. Although the rule has not yet been released, DHS has stated that development and implementation of the ammonium nitrate requirements would be resource-intensive and require trade-offs with ISCD's responsibilities under the CFATS program. For example, ISCD officials estimated that approximately 1,000 or more ammonium nitrate facilities would need to be inspected annually, which would likely increase ISCD's inspectors' workload since they would be responsible for both the CFATS program and the Ammonium Nitrate Security Program. According to ISCD officials, it will take at least 1 year from issuance of the final rule to establish the ammonium nitrate regulatory program. As a result, we could not assess the impact of the pending ammonium nitrate regulations on ISCD's approval of site security plans or on the CFATS program as a whole.

ISCD Is Conducting Compliance Inspections but Does Not Have Documented Processes and Procedures for Managing Facilities That Are Noncompliant with Security Plans

The CFATS regulation, consistent with the program's underlying statutory authority, authorizes ISCD to take enforcement action, such as issuing orders to assess civil penalties or to cease operations, against a covered chemical facility if, for example, a compliance inspection finds a facility to be noncompliant with its approved site security plan. The regulation also provides, however, that if a facility is found to be noncompliant, the facility shall be provided written notification, an opportunity for consultation, and time frames within which the facility is to ensure compliance. According to ISCD officials, based on the nature of the violations found thus far, it has been ISCD practice to exercise the discretion afforded to it under law and regulation and not to take enforcement actions but to instead work with noncompliant facilities on a case-by-case basis to ensure compliance. However, ISCD does not have documented processes and procedures to track facilities that are noncompliant with their approved site security plans and ensure facilities implement planned measures to become compliant.

ISCD standard operating procedures for inspections of covered facilities provide that inspectors are to report to ISCD, among other things, any recommended enforcement actions resulting from a compliance inspection.[53] The CFATS regulation also provides that if a facility is in violation of any part of the regulation, appropriate action may be taken, including the issuance of an order, compelling a facility to take actions necessary to become compliant.[54] For example, if a compliance inspection determines that a facility does not fully implement security measures as outlined in its site security plan, an order may be issued specifying actions the facility must take to remedy the instances of noncompliance, along with timeframes for coming into compliance. If a noncompliant facility does not comply with such an order, an order assessing a civil penalty of up to $25,000 per day for as long as the violation continues or, if warranted, an order to cease operations may be issued. ISCD officials stated that they consider an approved site security plan to be a contract between a facility and ISCD and a facility is therefore required to implement planned security measures by the deadlines specified in its site security plan—

commonly between 6 and 12 months after plan approval. According to ISCD officials, facilities that do not implement planned measures by these deadlines are noncompliant with their requirements under CFATS.

ISCD began conducting compliance inspections in September 2013 and as of April 2015 had conducted 83 compliance inspections out of 1,727 facilities with approved site security plans. Our analysis of these compliance inspections found that nearly half of facilities did not fully implement security measures needed to satisfy the risk-based performance standards—as required by the CFATS regulation—and therefore were not fully compliant with their approved site security plans. Specifically, 34 of 69 facilities that underwent compliance inspections and had completed compliance inspection reports as of February 2015 had not implemented one or more planned measures by deadlines specified in their approved plans.[55]

According to ISCD officials, ISCD has not exercised its authority to issue orders and take enforcement actions based on the nature of violations identified through compliance inspections it has conducted as of May 2015. Instead, ISCD officials told us they track noncompliant facilities individually and work with the facilities on a case-by-case basis to help ensure compliance. According to ISCD officials, as part of ISCD's review process following compliance inspections, officials track noncompliant facilities on a case-by-case basis using individual compliance inspection reports and do not close out a facility's report until issues of noncompliance are resolved. ISCD officials stated that, thus far, they have provided additional time to noncompliant facilities and conducted follow-on inspections to ensure implementation of planned measures. ISCD officials also stated that, thus far, it has been more productive to work with facilities on a case-by-case basis to ensure compliance than to use enforcement actions. However, we found that ISCD does not have documented processes and procedures for how officials and inspectors are to track noncompliant facilities and ensure that they take actions towards compliance when ISCD exercises its discretion not to take enforcement actions. According to ISCD officials, ISCD provided written guidance on this issue to inspectors in the March 2015 update to ISCD's inspections and approvals lessons-learned guidance. However, this guidance outlines requirements for the proper submission of compliance inspection reports by inspectors but does not provide documented guidance for how officials and inspectors are to ensure noncompliant facilities take actions to become compliant, such as how much additional time to provide to facilities to implement planned measures.

Our analysis of compliance inspections conducted by ISCD as of February 2015 indicated that inspectors made varying recommendations relating to the 34 facilities that inspectors found had not implemented planned measures as outlined in their site security plans by deadlines and therefore were not fully compliant with their approved site security plans:

- Inspectors recommended both a follow-on compliance inspection and an enforcement action for 1 of the 34 noncompliant facilities. However, ISCD officials later elected not to take an enforcement action in that instance and determined that they would instead work with the facility to ensure it implemented planned measures.
- Inspectors recommended a follow-on compliance inspection to verify the implementation of planned measures for 12 of the 34 noncompliant facilities.
- Inspectors did not recommend an enforcement action or follow-on compliance inspection for the remaining 21 of 34 noncompliant facilities.
- Inspectors reported that the 34 noncompliant facilities required additional time, beyond their compliance inspection date, to implement planned measures that were not implemented by deadlines in their site security plans. Among the 34 noncompliant facilities, additional time to implement these planned measures as intended ranged from 3 weeks to another year.

Figure 5 summarizes our analysis of compliance inspections conducted by ISCD.

Compliance inspections—after which ISCD makes compliance determinations to either mitigate or accept risk at facilities—are a critical stage in ISCD's risk management approach to implementing the CFATS program. The NIPP risk management framework states that risk assessment approaches should be documented, reproducible, and defensible—common principles that are broadly applicable to all parts of a risk methodology. Specifically, the NIPP calls for risk assessment approaches to clearly document what information is used, and that subjective judgments need to be transparent to minimize their impact and help ensure comparable results. In addition, *Standards for Internal Control in the Federal Government* states that all organizational transactions—which can include processes and procedures for dealing with regulated entities—need to be clearly documented and that this documentation should be readily available for examination in administrative policies or operating manuals.[56] According to ISCD officials, ISCD's processes and procedures—although not documented— track noncompliant facilities through the compliance process and ensure they implement planned measures. According to ISCD officials, ISCD's current method of tracking noncompliant facilities on a case-by-case basis using individual compliance inspection reports is sufficient. ISCD officials also stated that inspectors and ISCD headquarters officials evaluate the significance of security gaps and facilities' ability to implement planned measures when working to bring facilities into compliance.

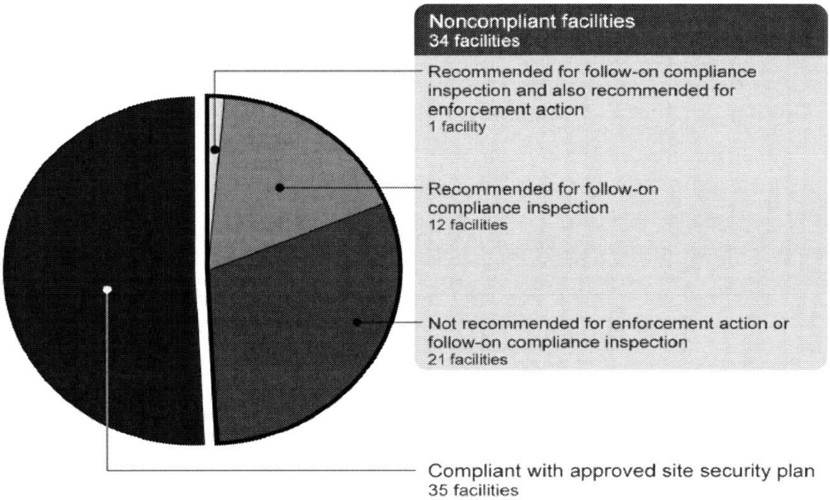

Source: GAO analysis of Infrastructure Security Compliance Division (ISCD) compliance inspection reports. | GAO-15-614.

Figure 5. Analysis of Chemical Facility Anti-Terrorism Standards (CFATS) Compliance Inspections Conducted as of February 2015.

By documenting its processes and procedures for tracking noncompliant facilities and ensuring planned measures are implemented, ISCD could better ensure consistency in how officials and inspectors address noncompliance in the CFATS program. Variations in the process for addressing noncompliant facilities—such as facilities having widely varying amounts of time to implement measures—may be warranted on a case-by-case basis. However, according to ISCD officials, facilities that do not implement planned security measures in accordance with their approved site security plans remain vulnerable to security threats until such measures are implemented. Such vulnerabilities could increase over time as the CFATS program matures and ISCD conducts compliance inspections for the approximately 2,900 additional facilities that, as of April 2015, had an approved site security plan or were assigned a final tier and awaiting approval of the site security plan. Documented processes and procedures that prescribe how to address noncompliance in lieu of taking enforcement actions, including the establishment of time frames to implement planned measures, would better position ISCD to achieve its broader mission of securing chemical facilities in a consistent and timely manner.

DHS's Performance Measure for CFATS Does Not Accurately Reflect Program Results

DHS's performance measure for the CFATS program, which was developed by ISCD and is intended to reflect security measures implemented by facilities and the overall impact of the CFATS regulation on facility security, does not solely capture security measures that are implemented by facilities and verified by ISCD.[57] Instead, DHS's performance measure reflects both existing security measures that were in place when facilities completed their site security plans and planned security measures approved by ISCD that facilities intend to implement within the fiscal year.[58] DHS reported this performance measure for the CFATS program in its *Annual Performance Report for Fiscal Years 2013- 2015* as the percent of performance standards implemented by the highest risk chemical facilities and verified by ISCD.[59] According to ISCD, the performance measure reflects the value of the CFATS program and its impact on reducing risk at facilities.

ISCD officials stated that they calculate the performance measure based on existing and planned measures identified in each facility's site security plan rather than based on the results of compliance inspections, at which ISCD verifies that facilities have implemented planned measures as intended. According to ISCD officials, they count the planned measure as having been implemented once the implementation date listed in a facility's site security plan has passed. However, according to ISCD officials, ISCD does not adjust the performance measure if it later determines that a facility did not implement a planned measure on time. In our analysis of compliance inspections conducted as of February 2015, we identified that 34 of 69 facilities did not implement one or more planned measures by intended deadlines. As a result, ISCD may have improperly counted some planned measures as implemented in the performance measure and therefore overstated the CFATS program's progress in reducing risk.

According to ISCD officials, when ISCD developed the performance measure, in fiscal year 2013, it had not yet begun conducting compliance inspections, at which ISCD verifies that facilities have implemented planned measures as intended. As a result, the performance measure does not take into account information gathered during compliance inspections, such as if facilities had implemented planned measures by the intended date. Since developing the performance measure, ISCD began conducting compliance inspections in September 2013, but ISCD officials stated that they did not begin verifying that planned measures had been implemented before including them in the performance measure at that point because they had conducted too few compliance inspections to produce meaningful performance data. ISCD officials also told us that the performance measure includes existing measures that were in place when facilities completed their site security plans because the goal of the performance measure is to capture the overall extent of security implemented by facilities. According to ISCD officials, it is difficult to assess the source of existing security measures because facilities could anticipate CFATS requirements and make enhancements prior to becoming a covered facility under CFATS. As a result, ISCD officials stated that it is a challenge to assess the baseline security measures already in place at facilities and therefore the impact of the CFATS program on improving those security measures. However, while it may be illustrative to report the overall extent of security implemented, including all existing measures does not reflect the value of the CFATS program and its impact on reducing risk at facilities, as ISCD's current performance measure is intended to do.

The NIPP calls for evaluating the effectiveness of risk management efforts by collecting performance data to assess progress in achieving identified outputs and outcomes. In addition, the purpose of CFATS, as stated in its regulation, is to enhance national security by furthering DHS's mission and lowering the risk posed by certain chemical facilities. Measuring the effectiveness of the CFATS program requires that facilities implement planned security measures identified as necessary to address vulnerabilities and that DHS evaluate implementation of these measures against CFATS performance standards. However, because ISCD's performance measure reflects both existing security measures that had not necessarily been implemented in response to CFATS and planned security measures that have not yet been verified as implemented, ISCD's performance measure does not reflect the value of the CFATS program and its impact on reducing risk at facilities, as stated in performance reports. As the CFATS program matures and ISCD conducts compliance inspections in greater numbers, revising current performance measures or adding new ones to accurately reflect only security measures that have been implemented and verified would help provide a more accurate picture of ISCD's progress and help ISCD ensure that the program is meeting its goals.

CONCLUSION

Individuals intent on using or gaining access to hazardous chemicals to carry out a terrorist attack continue to pose a threat to the security of chemical facilities and surrounding populations. DHS, through the CFATS program overseen by ISCD, has made progress in identifying chemical facilities that pose the greatest risks and in expediting the time it takes to approve security plans. However, DHS has not taken steps to mitigate errors in some facility-reported data and does not have reasonable assurance that it has identified all of the nation's highest-risk chemical facilities. Additionally, DHS cannot ensure consistency in how it addresses noncompliance in the CFATS program because it does not have documented processes and procedures. Finally, DHS's CFATS performance measure does not reflect security measures that facilities have implemented and that ISCD has verified, thus not accurately reflecting the value of the CFATS program and its impact on reducing risk at facilities.

RECOMMENDATIONS FOR EXECUTIVE ACTION

We recommend that the Secretary of Homeland Security direct the Under Secretary for NPPD, the Assistant Secretary for the Office of Infrastructure Protection, and the Director of ISCD to take the following two actions to ensure the accuracy of the data submitted by chemical facilities:

- provide milestone dates and a timeline for implementation of the new Top-Screen and ensure that changes to this Top-Screen mitigate errors in the Distance of Concern submitted by facilities, and

- in the interim, identify potentially miscategorized facilities with the potential to cause the greatest harm and verify the Distance of Concern these facilities report is accurate.

In addition, to better manage compliance among high-risk chemical facilities and demonstrate program results, we recommend the following two actions:

- develop documented processes and procedures to track noncompliant facilities and ensure they implement planned measures as outlined in their approved site security plans, and
- improve the measurement and reporting of the CFATS program performance by developing a performance measure that includes only planned measures that have been implemented and verified.

AGENCY COMMENTS

We provided a draft of this report to DHS for review and comment. DHS provided written comments, and technical comments, which we incorporated as appropriate. DHS concurred with all four recommendations and outlined steps that the National Protection and Programs Directorate (NPPD), Office of Infrastructure Protection (IP), Infrastructure Security Compliance Division (ISCD) will take to address them.

With respect to the first recommendation that NPPD provide milestone dates and a timeline for implementation of the new Top-Screen and ensure that changes to this Top-Screen mitigate errors in the Distance of Concern submitted by facilities, DHS noted that NPPD is developing a revised Top-Screen to eliminate the need for facilities to calculate and self-report Distances of Concern. In the interim, NPPD will verify the accuracy of Distances of Concern submitted in new Top-Screens. These actions, if fully implemented, should address the intent of the recommendation.

Regarding the first recommendation, DHS indicated in its letter that the impact from the erroneous Distance of Concern data is likely extremely minimal because only 85 of approximately 37,000 facilities that submitted Top-Screens were likely to have been assigned a lower risk status had they correctly reported their Distances of Concern, with potentially as few as two facilities actually having been impacted. However, as stated in our report, we based our estimates on a simple random sample and followed a probability procedure based on random selections. Our sample is only one of a large number of samples that we might have drawn, each of which could have provided different estimates. Therefore, the number of facilities affected by this issue may be as low as two or as high as 543.

DHS also stated that the erroneous Distance of Concern data impacts only preliminary tiering (categorization) results and additional information submitted through a security vulnerability assessment will affect the final tiering decision. As stated in our report, according to DHS data, 80 percent of facilities were determined not to be high-risk based upon information provided in the Top-Screen, which made those facilities not subject to additional requirements under the CFATS regulation. Thus, only a minority of facilities that submitted a Top-Screen were ultimately required to submit a security vulnerability assessment.

Regarding the second recommendation that NPPD identify potentially miscategorized facilities with the potential to cause the greatest harm and verify the Distance of Concern these facilities report is accurate, DHS stated that NPPD will review facilities that submitted Top-Screens with release-toxic chemicals of interest. NPPD will determine which facilities are most likely to potentially cause the greatest harm and will verify the Distances of Concern reported by the facilities. These actions, if fully implemented, should address the intent of the recommendation.

In response to the third recommendation that NPPD develop documented processes and procedures to track noncompliant facilities and ensure they implement planned measures, DHS stated that NPPD was in the process of developing and documenting such procedures. DHS also stated that NPPD has drafted requirements to update its case management system to separately track noncompliant facilities, a function not currently available in the system. According to DHS, in the interim, NPPD will monitor noncompliant facilities through a function in the case management system that keeps compliance inspection reports open until a facility implements all aspects of its site security plan. These actions, if fully implemented, should address the intent of the recommendation.

Finally, for the fourth recommendation that NPPD improve the measurement and reporting of the CFATS program performance by developing a performance measure that includes only planned measures that have been implemented and verified, DHS stated that NPPD will develop a performance measure that will provide an additional means to evaluate and illustrate the value of the CFATS program. According to DHS, the new measure will be included in ISCD's Fiscal Year 2016 Annual Operating Plan. These actions, if fully implemented, should address the intent of the recommendation.

Chris P. Currie
Director,
Homeland Security and Justice Issues

APPENDIX I: GAO SUMMARY OF DEPARTMENT OF HOMELAND SECURITY (DHS) PLANS TO IMPLEMENT THE PROTECTING AND SECURING CHEMICAL FACILITIES FROM TERRORIST ATTACKS ACT OF 2014

Required elements	DHS plans according to officials
Section 2. Chemical Facility Anti-Terrorism Standards Program	
The Secretary of Homeland Security's (Secretary) efforts to identify facilities, require the submission of Top-Screen and other information, establish risk-based performance standards, and require each covered facility to submit a security vulnerability assessment and to develop, submit and implement a site security plan. (6 U.S.C. § 622)	In August 2014, DHS's Infrastructure Security Compliance Division (ISCD) initiated a rulemaking process through an Advance Notice of Proposed Rulemaking to identify ways to makethe Chemical Facility Anti-Terrorism Standards (CFATS) program more effective in achieving its regulatory process. The rulemaking process comment period closed in October 2014, and ISCD is reviewing comments. According to ISCD officials, the rulemaking was included in the latest unified agenda, but it was listed as a long term item and thus did not have an expected timeframe associated with it.[a]

Required elements	DHS plans according to officials
Content of site security plans and employee input. (§ 622(b))	ISCD is considering requiring that facilities validate that employee input has been obtained to the greatest extent practicable in the development of their site security plans.
Approval and disapproval of site security plans, approval of alternative security programs, risk assessment policies and procedures for site security plan assessments. (§ 622(c)(1)-(3))	ISCD does not expect its review process to change.
Expedited approval program for site security plans, including guidance issued by the Secretary within 180 days of enactment (enacted December 18, 2014). (§ 622(c)(4))	The expedited approval process is the biggest change from the original statute, according to ISCD. DHS issued guidance for the expedited removal program in May 2015.
Audits and inspections by the Secretary, including (1) requirements for individuals working for nondepartmental or nongovernmental entities, who conduct audits or inspections, to report to a regional supervisor and (2) standards for training and retaining auditors and inspectors. (§ 622(d)(1))	ISCD is not planning to make any changes to its audits or inspections. Third party inspectors could be considered at a later date as appropriate, but ISCD did not have plans to use them as of June 2015. In terms of training, ISCD officials stated that in February 2015, they re-instituted a training task force to assess the effectiveness of ISCD's training efforts. As part of this effort, ISCD developed a statement of work to hire a contractor to analyze inspectors' training competencies.
The Secretary's establishment and carrying-out of a personnel surety program. (§ 622(d)(2))	ISCDs personnel surety program is awaiting Office of Management and Budget approval and ISCD did not have an estimate on when it will be finalized.

(Continued)

Required elements	DHS plans according to officials
Responsibilities of the Secretary to: 1. Consult with the heads of other federal agencies, states and political subdivisions, relevant business associations, and public and private labor organizations to identify all chemical facilities of interest; (§ 622(e)(1)) 2. Develop a security risk assessment approach and corresponding tiering methodology for covered chemical facilities; (§ 622(e)(2)) 3. Document each instance where tiering for a covered facility changes or a covered facility is no longer subject to the requirements; (§ 622(e)(3)) 4. Submit semiannual reports, beginning not later than 6 months after enactment, that includes for the period covered the number of covered facilities in the United States, changes in tiering, metrics addressing reviews and inspections, among other information. (§ 622(e)(4))	ISCD completed this requirement in response to Executive Order 13650 and is documented in the report of the Executive Order 13650 working group. ISCD intends to continue consulting with stakeholders as appropriate to identify chemical facilities of interest. In response to GAO recommendations to incorporate additional threat information into the risk tiering methodology, ISCD is updating how it addresses consequence, vulnerability and threat in its risk approach and corresponding tiering methodology. The risk methodology will be subject to a peer review when completed. ISCD currently documents each instance where a previously assigned tier changes and is modifying its case management system to facilitate required reporting on the number of such instances and the rationale for each. ISCD has drafted the first semi-annual report, which is currently in the DHS review and clearance process. ISCD is on schedule to submit the report to Congress in June 2015.

Required elements	DHS plans according to officials
Protection and sharing of information. (§ 623)	ISCD does not plan to make changes to the protection and sharing of information. ISCD is working to share permissible information with certain partners through Infrastructure Protection Gateway, a DHS portal containing tools and information to help partners prepare vulnerability assessments and risk analysis.
Civil enforcement, including providing notices of noncompliance and addressing circumstances of continued noncompliance; use of civil penalties and emergency orders. (§ 624)	ISCD has assessed the new statutory language and believes that some minor changes are necessary to make existing procedures conform to the new statutory language. These modifications are expected to be procedural in nature (i.e., do not require industry notification and an opportunity to comment) and ISCD is in the process of developing them.
Whistleblower protections, including: the establishment of procedures within 180 days of enactment for reporting violations to the Secretary. (§ 625)	Procedures for reporting violations already exist and are being expanded with the ability to submit reports by e-mail. In addition, ISCD is refining outreach materials to promote the availability of these mechanisms, and is working closely with the Department of Labor's Occupational Safety and Health Administration to establish mechanisms to enforce the whistleblower retaliation protections should they ever be invoked.

(Continued)

Required elements	DHS plans according to officials
The Secretary's efforts to promulgate regulations or amend the existing CFATS regulations to implement provisions of the Act and, within 30 days of enactment, whether the Secretary repealed any existing CFATS regulation that the Secretary determines is duplicative of, or conflicting with, the act. (§ 627)	This is a challenging deadline for DHS to meet, because repealingany of this without including any pending program changes resulting from the rulemaking process would not make sense. Therefore, as a matter of policy, DHS is going to miss these deadlines and continue to work through the rulemaking process.
Efforts undertaken by the Secretary, if any, to provide guidance and other support to "small covered chemical facilities" (as that term is defined in the Act) and whether the Secretary submitted the requisite report on best practices that may assist small covered chemical facilities in development of physical security best practices. (§ 628)	ISCD is developing guidance as part of its 18-month reporting requirement, but this may result in a separate deliverable to the small covered chemical facilities. No additional activities are anticipated.
Establish an outreach implementation plan to identify chemical facilities of interest and make available compliance assistance materials and information on education and training, not later than 90 days after the date of enactment of the act. (§ 629)	ISCD completed this plan on March 18, 2015, and shared it withfederal, state, local, commerce, and other stakeholders for comments in advance of completion.
Section 3. Assessments and Reporting: 1. The Secretary's commission of a third-party study to assess vulnerabilities of covered chemical facilities (using funds appropriated prior to enactment of the Act). (§ 3(b))	This third party assessment is under development. This report is under development.

Required elements	DHS plans according to officials
2. The Secretary's report, for submission to the Committee on Homeland Security and Governmental Affairs of the Senate and the Committee on Homeland Security and the Committee on Energy and Commerce of the House of Representatives not later than 18 month after enactment, on a. the CFATS program that includes certification by the Secretary that significant progress in identifying all chemical facilities of interest under 6 U.S.C § 622(e)(1) has been made; (§ 3(c)(1)(A)) b. certification by the Secretary that a risk assessment approach and corresponding tiered methodology under 6 U.S.C § 622(e)(2) has been developed; (§ 3(c)(1)(B)) c. the Secretary's assessment of implementation by DHS of recommendations by the Homeland Security Studies and Analysis Institute; (§ 3(c)(1)(C)) and d. a description of best practices that may assist small covered chemical facilities in the development of physical security best practices. (§ 3(c)(1)(D))	

Source: Public Law 113-254; interviews with ISCD officials. I GAO-15-614

Note: See Protecting and Securing Chemical Facilities from Terrorist Attacks Act of 2014, Pub. L. No. 113-254, 128 Stat. 2898 (2014); 6 U.S.C. §§ 621-29. Section 2 of the Act amends the Homeland Security Act of 2002, Pub. L. No. 107-296, 116 Stat. 2135 (2002), by adding Title XXI (§§ 2101-09)— Chemical Facility Anti-Terrorism Standards. All statutory references in this appendix are to the United States Code section, unless otherwise indicated.

[a] See 80 Fed. Reg. 35,030, 35,032 (June 18, 2015) (providing that DHS intends to publish a notice of proposed rulemaking by a date yet-to-be determined after considering public comments received on the advance notice of proposed rulemaking it issued soliciting comments for potential modifications to the CFATS program (79 Fed. Reg. 48,693 (Aug 18, 2014)).

APPENDIX II: SCOPE AND METHODOLOGY

This appendix provides details of our scope and methodology to answer each objective. For both objectives, we reviewed applicable laws,[1] regulations (including proposed rules),[2] Department of Homeland Security's (DHS) Chemical Facility Anti-Terrorism Standards (CFATS) program policies and procedures, and our prior reports on the CFATS program.[3] We also identified various criteria relevant to this program and compared the results of our analyses with these criteria, including the CFATS statute and rule, internal control standards,[4] project management guidance,[5] and policies and procedures outlined in the *National Infrastructure Protection Plan* (NIPP) risk management framework, which calls for risk assessments to be documented, reproducible, and defensible in order to generate results that can support investment, planning, and resource prioritization decisions.[6]

To address our first objective, on the extent to which the DHS's Infrastructure Security Compliance Division (ISCD) has identified and categorized facilities subject to the CFATS regulation, we reviewed laws applicable to facilities that possess chemicals, including the statutes authorizing and regulations governing the CFATS program and Executive Order 13650.[7] To address how ISCD has identified facilities that could potentially be subject to CFATS but have not yet self-identified or been identified by ISCD as being required to submit information to ISCD pursuant to CFATS, we reviewed and analyzed ISCD documentation related to its efforts to identify facilities, including documentation on coordinating with other federal agencies that regulate chemical facilities, and the Executive Order 13650 *Report for the President: Actions to Improve Chemical Facility Safety and Security—a Shared Commitment.*[8] To address how ISCD has categorized chemical facilities, we reviewed and analyzed documentation ISCD provides to facilities intended to guide them in submitting data that ISCD uses to preliminarily categorize facilities as high-risk, which renders them subject to additional requirements under the CFATS regulation during its preliminary screening process.[9] We reviewed ISCD applications and documents including web-based Chemical Security Assessment Tools (CSAT) applications—such as the Top-Screen—used to collect security information from facilities, the ISCD risk assessment approach used to determine a facility's preliminary risk tier, and policies and procedures on preliminary tiering. As part of this effort, we obtained and analyzed data submitted by facilities to ISCD as required by CFATS. First, to assess the reliability of data we obtained from CSAT, we reviewed system documentation, compared similar data sets for consistency, and interviewed knowledgeable ISCD officials about system controls and determined that CSAT data were sufficiently reliable for the purposes of this report. Second, we outlined the CFATS risk assessment approach and the three security issues upon which it is based—release-toxic/flammable/explosive, theft or diversion, and sabotage.

Finally, we analyzed a sample of ISCD data and compared it to criteria in the *National Infrastructure Protection Plan* (NIPP), *Supplemental Tool: Executing a Critical Infrastructure Risk Management Approach*, to determine the extent to which ISCD used facility-reported data to make categorization decisions that are documented and reproducible. Specifically, we selected a simple random sample of 475 facilities from the population of 36,811 facilities that submitted Top-Screens since the inception of the CFATS program in 2007 through January 2, 2015. During our review we identified 91 facilities from among the sample of 475 facilities that indicated that their facility had toxic release chemicals of interest above the screening threshold quantity. For these 91 facilities we tested the reliability of the Distance of Concern reported in Top-Screens. To determine how DHS intended respondents to calculate the Distance of Concern, we reviewed DHS guidance that directs facilities to use an online tool, RMP*Comp. We also reviewed instructions for using this tool. Using RMP*Comp, DHS guidance, and data from their most recently submitted Top-Screen (the type and quantity of chemical that could be released and terrain type surrounding facility), we calculated the Distance of Concern for the 91 facilities and compared these results to what facilities reported.

Because we followed a probability procedure based on random selections, our sample is only one of a large number of samples that we might have drawn. Since each sample could have provided different estimates, we express our confidence in the lower of our particular sample's results as a 95 percent confidence interval (e.g., plus or minus 7 percentage points). This is the interval that would contain the actual population value for 95 percent of the samples we could have drawn. For our determination of facilities potentially at risk of being missed by DHS because of errors in their RMP*Comp calculations we provide the 95 percent confidence interval associated with each estimate.

To corroborate and confirm our understanding of ISCD's approach to identify and categorize chemical facilities, we also interviewed ISCD officials knowledgeable about the processes to identify and categorize chemical facilities to corroborate our understanding of the program.

To address our second objective, on the extent to which DHS has approved site security plans, conducted compliance inspections, and measured results, we reviewed laws and regulations on how DHS is to approve site security plans and ensure compliance with the CFATS program. We analyzed ISCD data to identify the number of site security plans approved per month, and used the results of our analysis to estimate the number of months it could take ISCD to approve remaining site security plans. To assess the reliability of data we obtained from ISCD's case management system, we reviewed system documentation, conducted data and logic testing, and interviewed knowledgeable ISCD officials about system controls and determined that the data were sufficiently reliable for the purposes of this report. We note limitations in our analysis, including that our estimate does not take into consideration resource constraints ISCD may face as it implements the pending Ammonium Nitrate Security Program.[10] We analyzed 69 of 74 compliance inspection reports completed as of February 2015 to identify the extent to which the reports identify noncompliance among facilities that have reached the compliance inspection phase of the CFATS program. When we conducted our analysis, in February 2015, ISCD inspectors had not completed compliance inspection reports outlining the results of the inspections for 5 of the 74 facilities, so we did not include these facilities in our analysis. We compared ISCD's processes and procedures for tracking and monitoring noncompliant facilities to criteria in the NIPP for documented, reproducible, and defensible risk assessment approaches and *Standards for Internal Control in the Federal Government* for documenting transactions and significant decisions. To determine how DHS has measured results, we compared ISCD's performance measure with criteria in the NIPP for evaluating the effectiveness of risk management efforts by, among other things, collecting performance data to assess progress in achieving outputs and outcomes. We also interviewed ISCD headquarters officials regarding their processes and systems for authorizing, approving, and inspecting facilities for compliance. Finally, we conducted four site visits to observe (1) two authorization inspections and (2) two compliance inspections. We selected inspection locations in California, Oregon, Maryland, and Texas based on geographic dispersion and to cover different types of chemical facilities regulated by the CFATS program. While the information obtained from these inspections cannot be generalized to all inspections, it provides insight and context on how ISCD conducts CFATS inspections.

We conducted this performance audit from September 2014 to July 2015 in accordance with generally accepted government auditing standards. Those standards require that we plan and perform the audit to obtain sufficient, appropriate evidence to provide a reasonable basis for our findings and conclusions based on our audit objectives. We believe that the evidence obtained provides a reasonable basis for our findings and conclusions based on our audit objectives.

RELATED GAO PRODUCTS

Critical Infrastructure Protection: DHS Action Needed to Enhance Integration and Coordination of Vulnerability Assessment Efforts. GAO-14-507. Washington, D.C.: September 15, 2014.

Chemical Safety: Actions Needed to Improve Federal Oversight of Facilities with Ammonium Nitrate. GAO-14-274. Washington, D.C.: May 19, 2014.

Critical Infrastructure Protection: Observations on DHS Efforts to Implement and Manage Its Chemical Security Program. GAO-14-608T. Washington, D.C.: May 14, 2014.

Critical Infrastructure Protection: Observations on DHS Efforts to Identify, Prioritize, Assess, and Inspect Chemical Facilities. GAO-14-365T.

Washington, D.C.: February 27, 2014.

Critical Infrastructure Protection: DHS Efforts to Assess Chemical Security Risk and Gather Feedback on Facility Outreach Can Be Strengthened. GAO-13-353. Washington, D.C.: April 5, 2013.

Critical Infrastructure Protection: An Implementation Strategy Could Advance DHS's Coordination of Resilience Efforts across Ports and Other Infrastructure. GAO-13-11. Washington, D.C.: October 25, 2012.

Critical Infrastructure Protection: Summary of DHS Actions to Better Manage Its Chemical Security Program. GAO-12-1044T. Washington, D.C.: September 20, 2012.

Critical Infrastructure Protection: DHS Is Taking Action to Better Manage Its Chemical Security Program, but It Is Too Early to Assess Results. GAO-12-567T. Washington, D.C.: September 11, 2012.

Critical Infrastructure: DHS Needs to Refocus Its Efforts to Lead the Government Facilities Sector. GAO-12-852. Washington, D.C.: August 13, 2012.

Critical Infrastructure Protection: DHS Is Taking Action to Better Manage Its Chemical Security Program, but It Is Too Early to Assess Results. GAO-12-515T. Washington, D.C.: July 26, 2012.

Critical Infrastructure Protection: DHS Could Better Manage Security Surveys and Vulnerability Assessments. GAO-12-378. Washington, D.C.: May 31, 2012.

Critical Infrastructure Protection: DHS Has Taken Action Designed to Identify and Address Overlaps and Gaps in Critical Infrastructure Security Activities. GAO-11-537R. Washington, D.C.: May 19, 2011.

Critical Infrastructure Protection: DHS Efforts to Assess and Promote Resiliency Are Evolving but Program Management Could Be Strengthened. GAO-10-772. Washington, D.C.: September 23, 2010.

Critical Infrastructure Protection: Update to National Infrastructure Protection Plan Includes Increased Emphasis on Risk Management and Resilience. GAO-10-296. Washington, D.C.: March 5, 2010.

End Notes

[1] See 72 Fed. Reg. 17,792 (Apr. 9, 2007) (interim final rule) (codified as amended at 6 C.F.R. pt. 27); see also Pub. L. No. 109-295, § 550, 120 Stat. 1355, 1388-89 (2006).

[2] See Pub. L. No. 113-254, 128 Stat. 2898 (2014); 6 U.S.C. §§ 621-29. Specifically, the Act amended the Homeland Security Act of 2002, Pub. L. No. 107-296, 116 Stat. 2135 (2002), by adding Title XXI—Chemical Facility Anti-Terrorism Standards—and expressly repealed the program's authority under the fiscal year 2007 DHS appropriations act.

[3] GAO, *Critical Infrastructure Protection: DHS Is Taking Action to Better Manage Its Chemical Security Program, but It Is Too Early to Assess Results*, GAO-12-515T (Washington, D.C.: July 26, 2012).

[4] GAO, *Critical Infrastructure Protection: DHS Efforts to Assess Chemical Security Risk and Gather Feedback on Facility Outreach Can Be Strengthened*, GAO-13-353 (Washington, D.C.: Apr. 5, 2013).

[5] See DHS, *2013 National Infrastructure Protection Plan (NIPP), Partnering for Critical Infrastructure Security and Resilience* (Washington, D.C.: December 2013). The NIPP risk management framework is a planning methodology that outlines the processes for, among other things, setting goals and objectives; identifying critical infrastructure; assessing risk based on consequences, threats, and vulnerabilities; implementing protective programs and resiliency strategies; and measuring performance and taking corrective actions. Broadly defined, risk management is a process that helps policymakers assess risk, strategically allocate finite resources, and take actions under conditions of uncertainty.

[6] See Pub. L. No. 113-254, § 3(c)(2), 128 Stat. at 2918. The act further provides that GAO is to submit a second report not later than 1 year after the date of its initial report (roughly June 2016) that includes an assessment of the whistleblower protections provided in the act, and a third report not later than 1 year after GAO submits its second report that includes an assessment of the expedited approval program authorized under the act. We will summarize DHS officials' stated plans to implement provisions of the act in app. I of this report.

[7] 159 Cong. Rec. S1287, 1557 (daily ed. Mar. 11, 2013).

[8] Issued on August 1, 2013, Executive Order 13650—*Improving Chemical Safety and Security*—endeavors to improve chemical facility safety and security by facilitating coordination within and among all levels of government and with the owners and operators of such facilities. See Exec. Order No. 13650, 78 Fed. Reg. 48,029 (Aug. 7, 2013). Among other things, the order established a federal working group to improve federal coordination with state and local partners; enhance federal agency coordination and information sharing; modernize polices, regulations, and standards; and work with stakeholders to identify best practices.

[9] In May 2014, the working group established by Executive Order 13650 issued a report on its progress to date, findings and lessons learned, and next steps.

[10] See app. II for more information on the methodology used to perform this analysis. Facilities not preliminarily considered high risk are not subject to additional requirements under the CFATS regulation.

[11] Project Management Institute, Inc. *A Guide to the Project Management Body of Knowledge* (PMBOK® Guide), Fifth Edition (Newtown Square, Pennsylvania: 2013).

[12] In accordance with the DHS Appropriations Act, 2008, Pub. L. No. 110-161, Div. E, § 563, 121 Stat. 1844, 2083-90 (2007), DHS issued a proposed rule for the Ammonium Nitrate Security Program in August 2011. See 76 Fed. Reg. 46,908 (Aug. 3, 2011). As of July 2015, DHS had not implemented the program but, according to the DHS semiannual regulatory agenda published in June 2015, a final rule for the program is anticipated in October 2015. See 80 Fed. Reg. 35,030, 35,031 (June 18, 2015).

[13] GAO, *Standards for Internal Control in the Federal Government*, GAO/AIMD-00-21.3.1 (Washington, D.C.: November 1999).

[14] Pub. L. No. 109-295, § 550, 120 Stat. at 1388-89.

[15] 72 Fed. Reg. 17,688 (Apr. 9, 2007) (codified as amended at 6 C.F.R. pt. 27); 72 Fed. Reg. 65,396 (Nov. 20, 2007) (codified at 6 C.F.R. pt. 27, App. A). The interim final rule (i.e., the CFATS regulation), as subsequently amended, remains in effect. Appendix A has not been revised since its initial publication.

[16] Such facilities can include food-manufacturing facilities that use COI in the manufacturing process, universities that use the chemicals to do experiments, or warehouses that store ammonium nitrate, among others. Under the Protecting and Security Chemical Facilities from Terrorist Attacks Act, such a facility would be recognized as a "chemical facility of interest." See 6 U.S.C. § 621(2). Consistent with law and regulation, certain facilities—including, in general, facilities regulated under the Maritime Transportation Security Act of 2002 (Public Law 107-295), public water systems or wastewater treatment facilities, facilities owned and operated by the Department of Defense or the Department of Energy, and facilities subject to regulation by the Nuclear Regulatory Commission or in accordance with the Atomic Energy Act of 1954—are not subject to regulation under CFATS and are referred to as excluded facilities. See, e.g., 6 C.F.R. § 27.110(a).

[17] See generally 6 C.F.R. pt. 27, subpt. B.

[18] See Pub. L. No. 113-254, §§ 2, 4-5, 128 Stat. at 2898, 2918-19 (adding Title XXI— Chemical Facility Anti-Terrorism Standards—to the Homeland Security Act of 2002, Pub. L. No. 107-296, 116 Stat. 2135 (2002); 6 U.S.C. §§ 621-29)).

[19] Among other things, the act expands upon the regulatory definitions of "chemical facility" (defining "chemical facility of interest" as a facility, other than an excluded facility, that holds, or that the Secretary of Homeland Security has a reasonable basis to believe holds, a chemical of interest at a threshold quantity set pursuant to relevant risk-related security principles) and "covered chemical facility" (defining the term as a facility, other than an

excluded facility, that the Secretary identifies as a chemical facility of interest and, based upon review of the facility's information [i.e., the facility's Top-Screen], determines it meets the established risk criteria). See 6 U.S.C. §§ 621(2)-(4), 622(e). For purposes of this report, we refer to covered chemical facilities as being high-risk.

[20] See Pub. L. No. 110-161, Div. E, § 563, 121 Stat. 1844, 2083-90 (2007) (amending Title VIII of the Homeland Security Act of 2002 by adding Subtitle J—Secure Handling of Ammonium Nitrate); 6 U.S.C. §§ 488-88i.

[21] See 6 U.S.C. § 488a(a); see also 76 Fed. Reg. 46,908 (Aug. 3, 2011) (proposed rule). Consistent with the statute, DHS's proposed rule for the Ammonium Nitrate Security Program rule defines, among other things, "ammonium nitrate facility" as any person or entity that produces, sells, or otherwise transfers ownership of, or provides application services for, ammonium nitrate, as well as what constitutes ammonium nitrate for purposes of the program.

[22] The Terrorist Screening Database is the U.S. government's consolidated list of known and suspected terrorists.

[23] See 80 Fed. Reg. at 35,031 (indicating that a final rule for the program is anticipated in October 2015).

[24] For example, a facility that possesses butane at a quantity equal to or exceeding 10,000 pounds with a minimum concentration of 1.0 percent must submit information to DHS because the substance is considered flammable if subject to release. A facility that uses or maintains oxygen difluoride, however, must submit information to DHS if it possesses the substance at a quantity equal to or exceeding 15 pounds with a minimum concentration of 0.09 percent because it is considered vulnerable to theft for use as a weapon of mass effect. The "screening threshold quantity" is the quantity of a chemical of interest, upon which the facility's obligation to complete and submit the Top-Screen is based. See 6 C.F.R. § 27.105.

[25] Once a respondent submits the Distance of Concern, ISCD uses it to estimate the size of the area in which fatalities would occur and determines the population within that area.

[26] RMP*Comp was developed and is administered by the Environmental Protection Agency (EPA) to help entities regulated under the Clean Air Act with reporting requirements. As part of these reporting requirements, regulated entities must conduct an off-site consequence analysis to provide information about the potential consequences of an accidental chemical release. The off-site consequence analysis consists of two elements: (1) a worst-case release scenario, which evaluates the consequences of the release of the largest quantity of the regulated chemical that results in the greatest Distance of Concern, and (2) an alternative release scenario, which evaluates the consequences of a release event more likely to occur than the worst-case scenario but which could still affect the surrounding population.

[27] According to DHS data, 80 percent of facilities were not categorized as high-risk based upon information facilities provide in the Top-Screen.

[28] 6 C.F.R. § 27.215. Facilities categorized as Tier 4, however, may submit an Alternative Security Program (ASP) in lieu of the CSAT security vulnerability Assessment. See 6 C.F.R. § 27.235(a)(1). An ASP is a third-party or industry organization program; a local authority, state or federal government program; or any element or aspect thereof that has been determined to meet the requirements of and provide for an equivalent level of security to that established by the CFATS regulation. See 6 C.F.R. § 27.105. Facilities categorized as Tiers 1 to 3 may not submit an ASP in lieu of the security vulnerability assessment. 6 C.F.R. § 27.235(a)(2).

[29] 6 C.F.R. §§ 27.225, 27.235.

[30] The CFATS regulation establishes 18 risk-based performance standards that identify the areas for which a facility's security posture are to be examined, such as perimeter security, access control, and cybersecurity. See 6 C.F.R. § 27.230.

[31] According to ISCD officials, site security plans can also be sent back to facilities to be revised for any number of reasons. For example, during the preliminary review, if ISCD finds that a plan does not contain all the requisite data needed to meet regulatory requirements, ISCD can return the plan to the facility for more information.

[32] See 6 C.F.R. § 27.250.

[33] See 6 C.F.R. §§ 27.245(b), 27.250(b)(2), 27.300.

[34] 6 C.F.R. §27.300(a)-(b).

[35] According to ISCD officials, 32,777 chemical facilities—facilities that hold a COI at or above the screening threshold quantity—submitted a Top-Screen from 2007 to 2009; 1,899 facilities submitted a Top-Screen from 2010 to 2012 and 2,459 facilities have submitted a Top-Screen since 2013.

[36] GAO, *Chemical Safety: Actions Needed to Improve Federal Oversight of Facilities with Ammonium Nitrate,* GAO-14-274 (Washington, D.C.: May 19, 2014).

[37] ISCD's other efforts to identify potentially noncompliant chemical facilities include a hotline to report noncompliant facilities and outreach to industry associations and State Emergency Response Commissions.

[38] ISCD officials noted that because of differences in the data fields due to the different regulatory role of EPA, ISCD had to manually compare the EPA and CFATS facility data in spreadsheets, which resulted in ISCD identifying a high-number of facilities as potentially noncompliant that had already submitted a Top-Screen. Consistent with law and regulation, certain facilities, such as public water systems or wastewater treatment facilities, are not subject to regulation under CFATS and are not required to submit a Top-Screen. See, e.g., 6 C.F.R. § 27.110(b).

[39] As of April 2015, ISCD was still determining the risk of an additional 44 facilities, some of which may be ultimately categorized as high-risk.

[40] Reasons other states did not provide ISCD data varied. According to ISCD officials, some states responded that the information is publicly available and ISCD could retrieve it. Other states told ISCD they were unwilling to share data with ISCD. Some states did not respond to the request at all.

[41] The estimate from our statistical sample is 7,749 facilities. The associated 95 percent confidence interval is 6,409 to, 9,092 facilities.

[42] To calculate the Distance of Concern, DHS guidance directs facilities to use the RMP*Comp tool—a web-based tool on the EPA website—which allows users to enter data related to a chemical, such as the type and amount, and to calculate the magnitude of exposure if the chemicals were to be released.

[43] The estimate is 4,173 facilities. The associated 95 percent confidence interval is 2,798 to 5,822 facilities.

[44] The estimate is 2,384 facilities. The associated 95 percent confidence interval is 1,213 to 4,129 facilities.

[45] To determine this, ISCD officials told us that they calculated alternative Distances of Concern for facilities in our sample through the automated tool ISCD uses to categorize facilities' risk. Because we do not have access to this tool and it was beyond the scope of this engagement, we were unable to verify their results.

[46] The associated 95 percent confidence interval is 2 to 543 facilities.

[47] Project Management Institute, Inc. *A Guide to the Project Management Body of Knowledge* (PMBOK® Guide), Fifth Edition (Newtown Square, Pennsylvania: 2013).

[48] According to ISCD officials, ISCD approved an average of between 80 and 100 site security plans per month during calendar year 2014, and officials told us they believe they will maintain those approval rates moving forward. Using data in ISCD's case management system, we found that ISCD approved an average of 85 plans per month in 2014 and 90 plans per month over the last 6 months of the year. According to these data and ISCD's estimates, we calculated a range of estimates for the time needed by ISCD to resolve the backlog of unapproved plans assuming potential approval rates of 80, 90, and 100 plans per month.

[49] Our estimate does not include 270 site security plans that were actively in the review process as of April 2015, or 461 facilities that have yet to be assigned a final tier as of April 2015 and therefore may not all be required to complete a site security plan. In addition, as of January 2013, ISCD had assigned approximately 3,500 high-risk covered chemical facilities to a final tier. As of April 2015, there were approximately 2,900 covered chemical facilities with a final tier. According to ISCD officials, approximately 600 facilities are no longer covered chemical facilities subject to additional CFATS requirements.

[50] See GAO, *Critical Infrastructure Protection: DHS Efforts to Assess Chemical Security Risk and Gather Feedback on Facility Outreach Can Be Strengthened,* GAO-13-353 (Washington, D.C.: Apr. 5, 2013).

[51] In April 2013, we reported that ISCD revised its procedures for reviewing facilities' site security plans to address ISCD manager concerns that the original process was slow and overly complicated. See GAO-13-353. In May 2014, we reported that officials were taking other actions, such as updating the internal case management system, to reduce the time needed to review remaining site security plans. See GAO, *Critical Infrastructure Protection: Observations on DHS Efforts to Implement and Manage Its Chemical Security Program,* GAO-14-608T (Washington, D.C.: May 14, 2014).

[52] According to ISCD officials, ISCD has submitted the Information Collection Request for the Terrorist Screening Database requirement of the personnel surety standard to the Office of Management and Budget for review and clearance. In May 2014, we reported that, according to ISCD officials, once the personnel surety performance standard is finalized, ISCD plans to reexamine each approved plan to ensure that approved facilities are in compliance with the personnel surety performance standard. In the meantime, ISCD officials told us that ISCD is informing facilities in their letters of approval that they may be subject to additional review with regard to the Terrorist Screening Database requirement once the personnel surety performance standard is finalized. See GAO-14-608T. See also 6 U.S.C. § 622(d)(2) (requiring DHS to establish and carry out a Personnel Surety Program applicable to covered chemical facilities).

[53] Specifically, ISCD's standard operating procedures provides that the resulting compliance inspection report summary is to list all inconsistencies, identify follow-on actions for the facility or DHS, and categorize any recommended enforcement actions such as Failure to Allow Inspection (6 C.F.R. § 27.250); Insufficient Response/Failure to Respond to Information Request by the Assistant Secretary for Infrastructure Protection (6 C.F.R. § 27.200(a) or (b)); Security Measures Not Implemented in Accordance with Site Security Plans (6 C.F.R. § 27.225); Security Measures Not Implemented in Accordance with an Alternative Security Program (6 C.F.R. § 27.235); Maintenance of Records (6 C.F.R. § 27.255); and Improper Handling/Disclosure of Chemical-terrorism Vulnerability Information (CVI) (6 C.F.R. § 27 400(d)).

[54] See 6 C.F.R. § 27.300.

[55] When we conducted our analysis, in February 2015, ISCD had conducted compliance inspections at 74 facilities. However, ISCD inspectors had not completed compliance inspection reports outlining the results of the inspections for 5 of those facilities, so we did not include the facilities in our analysis.

[56] GAO/AIMD-00-21.3.1.

[57] In fiscal year 2013, ISCD updated the performance measure in DHS's *Annual Performance Report for Fiscal Years 2013-2015* to better reflect CFATS program progress. The measure calculates the percentage of risk-based performance standards (RBPS) that are confirmed through the SSP/ASP approval process as having been met by tier 1 and tier 2 covered facilities. ISCD created a separate measure and performance targets to track tier 3 and 4 facilities in fiscal year 2015.

[58] The numerator of ISCD's calculation for the performance measure is the total number of RBPS for tier 1 and 2 facilities that are satisfied by either existing measures or planned measures that are approved and intended to be completed within the fiscal year. To calculate the numerator, ISCD adds the number of RBPS with planned measures that are deemed necessary and are planned to be completed within the fiscal year, and the number of RBPS with existing measures only. The denominator of the measure is the total number of applicable RBPS for all final-tiered tier 1 and 2 facilities. The denominator is calculated by multiplying the total number of tier 1 and 2 facilities by 18 RBPS and then subtracting the RBPS that are not applicable to facilities based on their specific security concerns.

[59] ISCD officials stated that ISCD has additional performance measures intended to measure the impact of the CFATS program, such as the percent of facilities that include at least one planned measure in their site security plan and the number of facilities that are no longer considered high-risk because, for example, they eliminated chemical holdings. However, ISCD tracks these performance measures in annual operating plans and DHS does not report them for the CFATS program in its *Annual Performance Report for Fiscal Years 2013-2015*. According to ISCD officials, ISCD has also begun work to develop a new performance measure that would better identify the baseline level of security for chemical facilities and compare that against the level of security represented by facilities' approved site security plans. However, ISCD officials stated that development and analysis relating to this task have not been completed.

End Notes for Appendix I

[1] See, e.g., Department of Homeland Security Appropriations Act, 2007, Pub. L. No. 109- 295, § 550, 120 Stat. 1355, 1288 (2006) (requiring the issuance of interim final regulations establishing risk-based performance standards for security of chemical facilities and requiring vulnerability assessments and the development and implementation of site security plans for chemical facilities), repealed and supplanted by the Protecting and Securing Chemical Facilities from Terrorist Attacks Act of 2014, Pub. L. No. 113-254, 128 Stat. 2898.

[2] See, e.g., 6 C.F.R. pt. 27; *Ammonium Nitrate Security Program*, 76 Fed. Reg. 46,908 (Aug. 3, 2011) (proposed rule, to be codified at 6 C.F.R. pt. 31).

[3] For a list of prior GAO reports related to this issue, see the Related GAO Products list at the end of this report.

[4] GAO, *Standards for Internal Control in the Federal Government*, GAO/AIMD-00-21.3.1 (Washington, D.C.: November 1999). Internal control is an integral component of an organization's management that provides reasonable assurance that the following objectives are being achieved: effectiveness and efficiency of operations, reliability of financial reporting, and compliance with applicable laws and regulations. These standards, issued pursuant to the requirements of the Federal Managers' Financial Integrity Act of 1982 (FMFIA), provide the overall framework for establishing and maintaining internal control in the federal government.

[5] Project Management Institute, Inc. *A Guide to the Project Management Body of Knowledge* (PMBOK® Guide), Fifth Edition (Newtown Square, Pennsylvania: 2013). Standard project management practices include activities such as developing a timeline with milestone dates to identify points throughout the project to reassess efforts under way to determine whether project changes are necessary.

[6] DHS, *2013 National Infrastructure Protection Plan, Partnering to Enhance Protection and Resilience* (Washington, D.C.: December 2013).

[7] Issued on August 1, 2013, Executive Order 13650—*Improving Chemical Safety and Security*—endeavors to improve chemical facility safety and security by facilitating coordination within and among the all levels of government and with the owners and operators of such facilities. See Exec. Order No. 13650, 78 Fed. Reg. 48,029 (Aug. 7, 2013). Among other things, the order established a federal working group to improve federal coordination with state and local partners; enhance federal agency coordination and information sharing; modernize polices, regulations, and standards; and work with stakeholders to identify best practices.

[8] In May 2014, the working group established by Executive Order 13650 issued a report on its progress to date, findings and lessons learned, and next steps. Executive Order 13650 *Report for the President: Actions to Improve Chemical Facility Safety and Security—a Shared Commitment* (May 2014).

[9] In general, facilities not preliminarily considered high-risk are not subject to additional requirements of the CFATS regulation. If a facility makes certain changes, or if DHS determines the facility needs to submit a new Top-Screen, a facility may be required to submit additional information.

[10] In accordance with the DHS Appropriations Act, 2008, Pub. L. No. 110-161, Div. E, § 563, 121 Stat. 1844, 2083-90 (2007) (requiring that DHS issue a proposed rule not later than 6 months after enactment [enacted December 26, 2007] and a final rule not later than 1 year after enactment), DHS issued a proposed rule for the Ammonium Nitrate Security Program in August 2011. 76 Fed. Reg. 46,908 (Aug. 3, 2011). As of July 2015, DHS had not implemented the program but, according to the DHS semiannual regulatory agenda published in June 2015, a final rule for the program is anticipated in October 2015. See 80 Fed. Reg. 35,030, 35,031 (June 18, 2015).

INDEX

A

access, 51, 67
agencies, 29, 37, 38, 56, 60
ammonia, 39
ammonium, 28, 29, 32, 44, 65, 66
appropriations, 64
Appropriations Act, 21, 28, 29, 31, 32, 65, 69, 70
assessment, 1, 4, 5, 9, 18, 19, 22, 29, 33, 34, 41, 47, 53, 54, 58, 59, 61, 64, 66
assets, 33
audit(s), 4, 31, 55, 62
authority, 3, 5, 32, 46, 64, 66

B

benefits, 38

C

categorization, 30, 53, 61
certification, 5, 59
challenges, vii, 2, 3, 7, 10, 13, 15, 17, 20, 28, 29, 38
chemical(s), vii, 1, 2, 3, 4, 5, 6, 7, 8, 10, 19, 25, 26, 28, 29, 31, 32, 33, 34, 35, 37, 39, 40, 41, 43, 45, 49, 50, 51, 52, 53, 56, 58, 59, 60, 61, 62, 65, 66, 67, 68, 69, 70
chemical industry, 2
Clean Air Act, 66
commerce, 58
compliance, vii, 1, 2, 4, 5, 6, 7, 10, 14, 18, 19, 20, 21, 24, 25, 26, 27, 28, 29, 30, 34, 37, 44, 45, 46, 47, 48, 49, 50, 51, 52, 53, 58, 62, 68, 69, 70
conflict, 3, 5
Congress, vii, 1, 3, 5, 29, 56
Consensus, 22
constant rate, 13, 14, 16, 17
consulting, 56
coordination, 65, 70
covering, 3
credentials, 44
critical infrastructure, 64
cybersecurity, 67

D

data collection, 43
data set, 61
database, 26, 37, 39
deficiency(s), 4, 7, 34
Department of Agriculture, 37
Department of Defense, 65
Department of Energy, 65
Department of Homeland Security (DHS), v, vii, 1, 2, 3, 4, 5, 6, 7, 8, 9, 10, 11, 12, 13, 14, 15, 16, 17, 18, 19, 20, 21, 22, 23,

24, 25, 26, 27, 28, 29, 30, 31, 32, 36, 39, 40, 44, 49, 50, 51, 52, 53, 54, 55, 56, 57, 58, 59, 60, 61, 62, 63, 64, 65, 66, 67, 68, 69, 70
Department of Labor, 57
Department of Transportation, 37
dispersion, 31, 62
draft, 52

E

education, 58
e-mail, 57
emergency, 57
enforcement, 34, 45, 46, 47, 49, 57, 68
environment, vii, 3
Environmental Protection Agency (EPA), 27, 37, 39, 66, 67
evidence, 31, 63
Executive Order, 29, 37, 56, 60, 65, 70
exercise(s), 45, 46
exposure, 26, 33, 39, 67

F

federal agency, 65, 70
federal authorities, 32
federal government, 30, 47, 62, 65, 66, 70
Federal Register, 22, 23, 24
financial, 70
fiscal year, 49, 50, 64, 69
food, 65
force, 3, 55
funds, 58

G

GAO, 6, 17, 18, 23, 25, 26, 27, 29, 36, 38, 40, 42, 48, 54, 56, 59, 63, 64, 65, 67, 68, 69, 70
GPRA, 27
guidance, 26, 27, 30, 35, 39, 43, 46, 55, 58, 60, 61, 67
guidelines, 27

H

height, 22
historical overview, 2
Homeland Security Act, 31, 59, 64, 65, 66
House, 21, 22, 23, 59
House of Representatives, 59

I

identification, 33, 37
improvements, 20, 43
individuals, 32, 55
industrial sectors, 3
industry(s), 2, 5, 57, 66, 67
information exchange, 5
information sharing, 65, 70
infrastructure, 31
ingredients, 28
injury, 26, 33, 39
inspections, vii, 2, 3, 6, 7, 14, 15, 18, 19, 20, 21, 25, 27, 29, 30, 43, 45, 46, 47, 49, 50, 51, 55, 56, 62, 69
inspectors, 20, 35, 43, 44, 45, 46, 47, 48, 55, 62, 69
interface, 21
investment, 38, 60
issues, 3, 4, 6, 7, 10, 19, 20, 33, 35, 44, 46, 61
iteration, 40

L

laws, 25, 29, 30, 60, 62, 70
laws and regulations, 29, 30, 62, 70
legislative proposals, 3
light, 44

M

magnitude, 35, 67
management, 27, 30, 35, 41, 43, 47, 53, 56, 60, 62, 64, 68, 70

manufacturing, 2, 65
Maritime Transportation Security Act, 65
Maryland, 26, 31, 62
mass, 25, 28, 66
materials, 22, 57, 58
matter, 58
measurement, 52, 53
methodology, 31, 41, 47, 56, 59, 60, 64, 65
mission, 49, 50
modifications, 3, 7, 57, 59

N

national security, 50
Nuclear Regulatory Commission, 65

O

Office of Management and Budget, 21, 55, 68
Office of the Inspector General, 21
officials, 25, 27, 30, 35, 37, 39, 41, 43, 44, 45, 46, 47, 48, 49, 50, 54, 55, 56, 57, 58, 59, 61, 62, 64, 67, 68, 69
oil, 28
Oklahoma, 28
operations, 31, 35, 45, 70
opportunities, 28
outreach, 57, 58, 67
oversight, 10, 40
ownership, 66
oxygen, 66

P

parallel, 15, 20
peer review, 29, 56
penalties, 45, 57
petroleum, 10
planned action, 43
policy, 58
policymakers, vii, 2, 3, 64
population, 25, 33, 61, 66
President, 29, 60, 70
principles, 47, 65
probability, 53, 61
project, 20, 30, 41, 60, 70
protection, 57
public health, vii, 3

R

radius, 33, 39
recommendations, 26, 29, 47, 52, 56, 59
reforms, 18
regulations, 1, 3, 5, 25, 31, 44, 58, 60, 65, 69, 70
regulatory framework, 32
regulatory requirements, 2, 28, 34, 39, 67
reliability, 25, 30, 60, 61, 62, 70
requirement(s), 2, 4, 5, 6, 10, 21, 22, 28, 30, 31, 33, 34, 35, 37, 38, 39, 41, 44, 46, 50, 53, 55, 56, 58, 60, 65, 66, 68, 70
resources, 13, 64
response, 29, 37, 50, 53, 56
retaliation, 57
risk(s), vii, 1, 2, 3, 4, 5, 6, 7, 8, 9, 10, 11, 17, 19, 24, 25, 26, 28, 29, 30, 31, 33, 34, 35, 37, 39, 40, 41, 44, 46, 47, 49, 50, 51, 52, 53, 54, 55, 56, 57, 59, 60, 61, 62, 64, 65, 66, 67, 68, 69, 70
risk assessment, 26, 29, 30, 33, 39, 40, 41, 47, 55, 56, 59, 60, 62
risk management, 30, 41, 47, 50, 60, 62, 64
rules, 60

S

sabotage, 33, 61
safety, 65, 70
scope, 31, 60, 67
Secretary of Homeland Security, 4, 51, 54, 65
security, vii, 1, 2, 3, 4, 5, 6, 7, 9, 10, 11, 12, 13, 14, 15, 16, 17, 18, 21, 22, 23, 25, 26, 28, 29, 30, 31, 33, 34, 41, 42, 43, 44, 45, 46, 47, 48, 49, 50, 51, 52, 53, 54, 55, 56, 58, 59, 60, 62, 65, 66, 67, 68, 69, 70

security threats, 48
Senate, 21, 23, 59
September 11, 3, 23, 63
services, 66
spreadsheets, 67
stakeholders, 56, 58, 65, 70
state(s), 2, 19, 21, 37, 38, 41, 47, 56, 58, 65, 66, 67, 70
statutes, 60
statutory authority, 3, 45
storage, 10, 24
structure, 5
subjective judgments, 47
supervisor, 55

T

tanks, 10, 24
target, 19, 41
teams, 35, 43
technical comments, 52
terrorism, 27, 32, 68
terrorist attack, 4, 33, 51
terrorists, 28, 66
testing, 62
theft, 32, 33, 61, 66
threats, 41, 64
threshold level, 8
time frame, 20, 45, 49
time periods, 19
Title V, 66
tracks, 69
trade, 44
trade-off, 44
training, 55, 58
transactions, 30, 47, 62
treatment, 65, 67

U

United States, v, 2, 25, 28, 56, 59
universe, 2, 8
universities, 65
updating, 56, 68

V

variations, 27
vulnerability, 1, 3, 4, 5, 9, 10, 18, 22, 29, 31, 33, 34, 53, 54, 56, 57, 66, 69

W

Washington, 63, 64, 65, 67, 68, 70
wastewater, 65, 67
water, 33, 65, 67
weapons, 28
web, 26, 32, 39, 60, 67
White House, 21
workers, 2
workload, 44